BACKROADS & BYWAYS OF
MARYLAND

Turkey Point Lighthouse, on a bluff at the head of the Chesapeake Bay in Elk Neck State Park

BACKROADS & BYWAYS OF
MARYLAND

Drives, Day Trips
& Weekend Excursions

Leslie Atkins

The Countryman Press
Woodstock, Vermont

We welcome your comments and suggestions.

Please contact

Editor
The Countryman Press
P.O. Box 748
Woodstock, VT 05091

or e-mail countrymanpress@wwnorton.com.

Backroads & Byways of Maryland
ISBN 978-0-88150-926-7

Book design by Hespenheide Design
Map by Erin Greb Cartography, © The Countryman Press
Interior photos by the author
Composition by Chelsea Cloeter

Published by The Countryman Press, P.O. Box 748, Woodstock, VT 05091

Distributed by W. W. Norton & Company, Inc., 500 Fifth Avenue, New York, NY 10110

Printed in the United States of America

10 9 8 7 6 5 4 3 2 1

To my niece, Rebecca, I dedicate this book with my love and the hope that she continues the tradition of creativity in our family. And I dedicate this book to my brother, Richard, who captures his memories of our family and our lives in Maryland through his music and plays. Richard and I express ourselves differently, yet our desire to capture those special Maryland moments is the same.

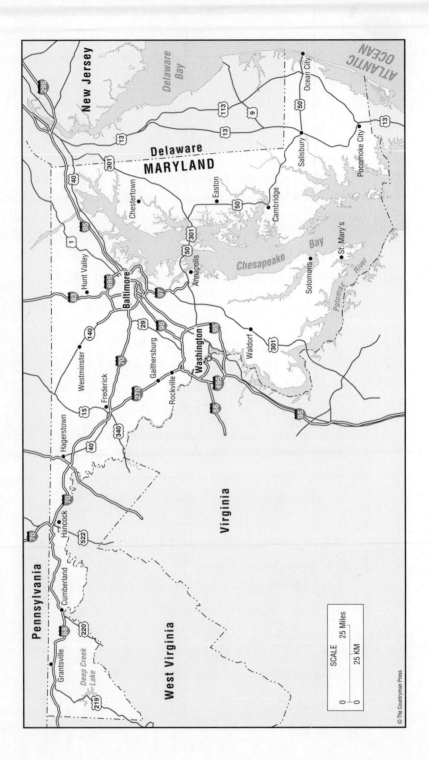

Contents

Acknowledgments

While an author gets a lot of attention, a book like this one is, in many ways, a group project. I would like to thank first and foremost the wonderful experts at Countryman Press, especially Kim Grant, Lisa Sacks, Kermit Hummel, Rosalie Wieder, Doug Yeager, and Tom Haushalter—my heartfelt appreciation goes out to them for their wise counsel, encouragement, and kindnesses.

I also appreciate the role of my parents, Esther and Arthur Atkins, who raised me in Maryland and taught me the joy of discovery. Dad always loved to take road trips and Mom loved to stop at roadside stands to buy fresh peaches or apple pie, which my brother Richard and I ate from the backseat. That was an idyllic time, when as a family we took unfamiliar roads to see what we would find.

My grandparents and great-grandparents came to America through the port of Baltimore to make their home here after traveling halfway around the world. It is not easy to leave the country of your birth, learn a new language, and adopt a new culture. I admire my grandparents and great-grandparents for that reason and for many others. I inherited my family's gene for exploration and the willingness to take risks. But my exploration of unfamiliar and isolated places in Maryland is tame in comparison to the journeys made by my ancestors.

I wish to thank all the people who welcomed me into their homes and hearts while I was researching this book. I was lucky to enjoy the assistance of many who opened doors and revealed vistas everywhere I went. These people are too numerous to mention, but I appreciate the help of each and every person who assisted me in uncovering the nooks and crannies throughout the state.

To my neighbors in Maryland who welcomed me with open arms, love, and friendship, I will always be grateful. And to the many strangers whom I met, the ones who shared their knowledge of local history and suggestions for places to see, you are the reason I love to travel, for it brings out the best in those you meet along the way. In the process, it brings out the best in me too.

Introduction

Maryland is an easy place for everyone to feel at home—whether they've spent most of their lives here as I have, or whether they're visiting for the first time.

A sublimely drivable state with well-kept roads and backroads, Maryland brims with charm and history. Because of its ethnic diversity and varied geography, Maryland is sometimes described as America in miniature. That characterization is misleading, however, for though the state has a wide variety of attributes, it is unique unto itself.

I was born and raised in Baltimore, the center of my world as a child and the state's largest city. Now that I've traveled extensively outside Maryland's boundaries and those of the entire United States, I've come to love Maryland anew, for it holds up well under scrutiny and compares positively to the many other places I've traveled.

When you're raised somewhere, it can seem old hat. But not so Maryland, which maintains its historic roots (and routes) while continually reinventing itself to keep up with the times. I'm sentimental because this is my home, but dispassionately speaking, Maryland is a must-see and a must-do for those who love traversing backroads, for water lovers who enjoy fishing and boating or just looking out upon large bodies of water, for those who love great food, and for those with a penchant for American history.

Maryland certainly leaves an indelible impression. There's so much to see that you can return many times without duplicating any single adven-

Canada geese on their way to US 50

ture—whether watching a baseball game, fishing along the rivers and streams, taking in the lighthouses on the Chesapeake Bay, hiking in the mountains, biking on the Ocean City boardwalk, or walking the side streets of Baltimore's neighborhoods enjoying many hidden and historic attributes.

Transportation routes—by water, rail, and road—helped define Maryland, and it is to those we turn in exploring the state. There are old mill towns, Civil War battlefields and scars, and tremendous shopping opportunities for those with a desire for antiques as well as those who like their acquisitions new. There are tremendous recreational activities, many of which are water-related. Sports are an integral part of life in the state, both the participatory and spectator varieties. Fans of baseball, football, and thoroughbred racing, in particular, will not be disappointed.

One cannot ignore Maryland's closeness to its waterways—the Atlantic Ocean, the rivers, and the Chesapeake Bay, which permeate life in this state where fishermen and crabbers influence local culture. Yet the state's immigrant roots also create an affinity toward European styles and ways, helping to make Maryland unique.

Among its other characteristics, Maryland is also a beach state, with the well-loved Ocean City and its extensive boardwalk leading to riches of seashore fun.

Literature, film, and television have captured the state's appeal, and portrayals of Maryland are now woven into the fabric of popular culture. Maryland is actually filled with an assortment of real characters and intriguing characteristics, making visits ever interesting and always provocative.

Perhaps the best part of traveling in Maryland is uncovering the little-known yet appealing country store, vista, or historic marker—not secret exactly, but certainly obscure. Discovery, after all, is one of the greatest factors in the desire to explore. Maryland is ripe for such exploration and it seems the more you uncover, the more you have yet to find.

It takes time and extra effort to find out-of-the-way places and talk to people you would not normally meet. But it is the extra effort that makes discovery special, for the journey is always as important as the destination.

When traveling along backroads and byways, one never knows what intriguing barn or creek will be just around the bend. A roadside stand, thoroughbred horses grazing in a field, or a pretty view can provide immeasurable pleasure.

To gain the most enjoyment, build flexibility into your schedule so you have time to pull over and watch something unexpected, snoop around an antiques store, or indulge in a dish of homemade ice cream. It is those moments that children most remember and adults truly treasure.

The abundance of riches throughout Maryland makes it virtually impossible to write about every nook and cranny, every village and marina. This book highlights many places and activities, but there are many more wonderful places for which there just wasn't room. However, those that are covered represent the diversity of landscapes and ways of life in Maryland.

I met some guests at a small-town inn who expressed it this way: "We just ramble. We know how many days we can be gone and we'll see a sign and say, let's go that way. We don't like schedules and we always loop back through the mountains. Some people love the beach. We like the mountains. We have a general idea which direction we want to go, and we'll see someplace that looks good and we'll go there."

Take your time, wander around, savor the experiences, enjoy the sights and sounds and smells and tastes…Maryland is wonderful. I know you'll love it.

Leaving Ewell for return to the mainland

1

Crisfield and Smith Island

Discovering Isolation on the Eastern Shore

Estimated length: 70 miles by car plus 30 (nautical) miles by boat
Estimated time: Overnight or weekend

Getting there: From Salisbury, take US 13 south approximately 20 miles. Bear right onto MD 413 south and drive another 14 miles to Crisfield. (Going south into Crisfield, MD 413 is Maryland Avenue and then becomes West Main Street at 7th Street, but it is still MD 413.) At the public City Dock, meet the local ferry captain, Larry, and climb aboard the *Captain Jason II,* where you'll stow your bag in the covered hatch on deck and ride to Smith Island with locals who came to the mainland on errands. (Park your car at the municipal lot across from the dock; it's free during the day and only $5 per night with an advance parking permit from the local police department or county tourism office.)

Highlights: Maryland's only inhabited offshore island group. Boat ride in Tangier Sound, part of the Chesapeake Bay. Peaceful sojourn at a bed & breakfast seemingly in the middle of nowhere. Biking and kayaking. Smith Island cake—watching it made and enjoying the result. Meeting watermen and their families. Town of Crisfield.

Part of this trip you'll be driving to **Crisfield** from **Salisbury** on the Eastern Shore. But you'll need to stow away your car keys for a while because you're visiting an island that's 12 nautical miles from the mainland. There is no bridge across those 12 miles (further north, the Chesapeake Bay Bridge is only 4.3 miles long). And there are no airports.

Once you reach Crisfield, you will park your car. Then you will rely upon the casual yet efficient island ferry system that services Smith Island. Except for the Smith Island Cruise boat, these are not large ferries in the usual sense; instead they are regular motorboats used for mail and freight and for transporting locals. Going overnight to **Tylerton**—the most remote part of **Smith Island**—you will travel on one of the smaller boats with few, if any, other visitors.

By its very isolation, Smith Island has evolved into a teeny tiny world unto itself and unlike any other. It is decidedly Maryland, since life there revolves around the Chesapeake Bay and its bounty—crabbing, oystering, and fishing. Though it is isolated like islands elsewhere, it has little similarity to places in the Caribbean or South Pacific. It is also unlike remote areas in Maine or Florida or South Carolina. Smith Island almost defies definition, for there is nothing else quite like it.

The minute you are on the boat heading out of Crisfield, it becomes apparent that this is a totally different kind of excursion than any other you'll experience. The captain obviously knows these waters like the back of his hand. There may be fresh produce on the deck in cardboard boxes and a screen door or love seat strapped to the boat's roof.

The Smith Islanders who take the boat back and forth talk to one another in their unusual accents and shorthand jargon about events that matter in their lives. It seems as if you are part of a play or film, watching and eavesdropping on a world that's different from your own. Smith Islanders are still speaking English, they still live in Maryland, and they are American. But they are self-sufficient like few others, know one another's families and stories, and are apparently comfortable with strangers who ask them a lot of questions.

There's no government and no police on the island. That may sound like anarchy, but after generations of living together, the Islanders have learned to help each other through adversity and just everyday life in a remote, isolated place. They do have disagreements but they have a practiced ability to work them out without undue emotion. There's also the Methodist churches on the island, which are the center of much social and spiritual activity.

Smith Island is actually two islands with three villages—**Ewell** (pro-

The Inn of Silent Music across the road from Tyler Creek and the Chesapeake Bay

nounced "Yule") and **Rhodes Point** on one shared island and Tylerton on the other. Smith Island was settled by the British in the 1600s and they later used it as a base of operations during the Revolutionary War. There are only about two hundred permanent residents throughout the three villages combined, plus some day-trippers who visit Ewell for a few hours on the tour boats. There are also a few overnighters (as you will be), plus occasional renters and summer "foreigners."

You'll want to stay at the **Inn of Silent Music** in Tylerton. There are only three rooms for rent at this small but delightful bed & breakfast, so advance reservations are an absolute must. Once there, it won't seem odd that so few rooms are available in Tylerton, for this is as quiet a place as you can imagine. There are some roads, but the only vehicles are a few pickup

Map of the island on the side of Tylerton's pump house

trucks used to haul heavy items around. After all, there's hardly anywhere to drive—the island is so small.

Everyone walks or bikes or rides around in electric golf carts until the carts lose their electrical charges. In fact, that's one of the favorite pastimes of residents here—riding round and round with friends in a golf cart. One woman even laughs that she is "Driving Miss Daisy," in reference to her more elderly friend and golf cart companion. Young girls also enjoy the diversion; instead of cruising in a car, they go "cruising" in a golf cart.

When you arrive in Tylerton, someone from the inn will meet you at the dock and help with your luggage. (Don't pack too much; everything is easier here if you travel light. However, do bring bug spray, suntan lotion, and sunglasses; you will need them.) On your way to the inn, which is only a few blocks from the dock, you'll see a map of the island painted on the side of the pump house, which supplies the village with fresh water. There are also numerous crab shanties plus docks with small workboats, but very few people.

The inn is a B&B in a house right on a marshy peninsula of land. The house fronts Tyler Creek, beyond which is the main channel of the Chesapeake Bay, so it is quite dramatic. There's a spacious screened-in porch on the front of the house where much of the activity of the B&B takes place. Meals are served at the table on the screened-in porch unless there's a storm and then it's inside to the large, comfortable kitchen.

You'll get settled in your room and then you'll have a choice of activities. You can take a walk around—there is not far to go—and you can stop in at the women's crab-picking co-op to learn about this traditionally labor-intensive work. On your walk around, you will probably even see a pet peacock that lives on the island, as well as numerous cats. You can also borrow a bike or a kayak from the B&B and get some exercise. Or you can read, talk, sleep, or do nothing but watch the birds and the weather and the workboats and relax.

Birders and even novices will marvel at the herons, egrets, ospreys, gulls, and numerous waterfowl. Anyone who likes quiet, or desires it, will find this a uniquely intriguing interlude. You will hear water lapping against a retaining wall, the breeze rustling the marsh grasses, and the sounds of birds, but little else except perhaps an occasional boat motor or thunderstorm.

This out-of-the-way destination is a hidden treasure, hard to reach and untouched by some, but not all, modern conveniences; yet it is still comfortable. Make sure to talk with Smith Islanders; with few entertainments, they amuse themselves by becoming great storytellers. Their stories are mixed with lots of humor. "Be careful if you're talking about someone around here," one warns, smiling. "You're probably talking to their cousin."

These teeny tiny islands in the southernmost part of Maryland, a distance of 12 nautical miles from the mainland, offer a glimpse into the traditional lives of watermen and their families, as well as a quiet, peaceful, idyllic existence. Not that life here isn't hard, for the physical work these families do in and around the water is backbreaking and laborious. Yet it is a way of life that has sustained generations and continues in this small place among a few families.

Unlike most other B&Bs, Inn of Silent Music serves dinner, and it is the only place in Tylerton to do so. Luckily, innkeeper Linda Kellogg is a gourmet chef, so you are in for a treat. She will accommodate allergies and preferences, so give her this information in advance. Her husband and co-innkeeper Rob Kellogg assists, and both sit down to have dinner with their guests. The Kelloggs are interesting and entertaining, so dinner is a pleasure.

The B&B does not provide wireless Internet, and cell phone reception is spotty at best. Be prepared to disengage and treasure the remoteness.

In the morning, you'll have another gourmet meal at breakfast, again on the front porch. Then you have the same activities as yesterday from which to choose. Or you can have the innkeepers contact Captain Waverly, who runs a water taxi service from Tylerton to Ewell on his small motorboat; Ewell is on the larger island that makes up the other part of Smith Island.

Ewell is a little bit larger than Tylerton, and there's slightly more to see there. In fact, after just one day and night spent in Tylerton, Ewell may seem almost commercial in comparison, though it's really not. Take Captain Waverly's water taxi there so you can have a crabcake lunch at Bayside Inn—their crabcakes are superb. You should also stop in the Visitor's Center and Cultural Museum to watch the 20-minute video—it's both humorous and provocative. Then walk or rent a golf cart to go see the women baking Smith Island cakes at **Smith Island Baking Company**.

Traditionally, Smith Islanders have lived off the Bay's riches—crabs and oysters in particular. With the Bay's bounty diminished due to environmental and other causes, the women on Smith Island have taken to commercializing their long tradition of baking ten (or more)-layer cakes with chocolate fudge icing between each layer.

In Ewell, the Smith Baking Company is located in a former hardware store and it helps employ islanders. It also helps feed enthusiastic tourists who visit and others who order the cakes online or by phone for shipment via the tiny post office across the street from the bakery. The bakery sells the cakes in white paper boxes or cheerful reusable yellow cake tins. Most of the work is done in the morning, but they're open until 3 PM to sell to walk-in visitors from the tour boats.

Among the women who work in the bakery, many have other jobs as well. A few pick crabs, one cuts grass…and so it goes in this isolated world where a few extra dollars can come in handy when the seafood catch is less than hoped for.

In Ewell, you will see a few cars and pickup trucks, but that is not the primary means of getting around. As on the rest of Smith Island, transportation is mostly by foot, by boat, by bike, by canoe or kayak, and by electric golf cart. There are ambulances and fire trucks, but no police or sheriffs are stationed on the island. There's no mail delivery either. Instead, residents go to the little post offices (in each community) to pick up their letters.

At the post office in Ewell, for instance, the boxes are mounted on the counter. They are obviously old. When questioned, the postmistress explains that they have combination locks, but most of them don't work. She either hands the mail to people when they come in or some people just leave their boxes unlocked so they can take their own mail. Throughout the communities that make up Smith Island, people don't generally lock their doors either. It's so small that everyone knows everyone else's business, and they all pitch in to help out if there's a need. Thus, it's like everybody is family.

How long your visit lasts will help determine when and where you catch a boat back, either first to Tylerton from Ewell or else directly to Crisfield from Ewell. The hosts at the inn will help make the arrangements

SMITH ISLAND CAKE

The state of Maryland is proud of its local specialties. The Diamondback Terrapin is the state reptile. The black-eyed Susan is the state flower. And the state decided in 2008 to make the Smith Island cake the state dessert. It didn't matter that Smith Island is a tiny spot of land in the middle of the bay at Maryland's southernmost border. It didn't matter that only a few hundred people live there. It didn't matter that some Smith Island cakes are 10 layers and some are 12.

Smith Island cake goes back generations and its origins are fuzzy. It is really a cottage industry, but recently a commercial operation started in Ewell. Legend has it that the watermen's wives sent the cakes with their husbands on the boats. Making the cakes with multiple thin layers and using fudge for frosting allowed them to stay moist much longer—important for men who stayed out on the water for days at a time. Now it's a more widely known dessert; and of course, it's great fun to eat a slice of cake with 8 or 10 or 12 layers.

Among Smith Islanders, some cakes are made with doctored cake mixes; others are made from scratch. The commercial bakery makes them from scratch but both versions are good. The traditional version is yellow cake with chocolate fudge. Less traditional versions use different kinds of cake, fruit, and glazes, but the layers are always thin—a source of pride when they are uniform.

for you. The islanders know the boat captains and their schedules by heart, and though it seems rather casual to newcomers, it works. Just make sure you ask which dock to wait on, as there are several on each island; otherwise you might miss your prearranged ride by waiting in the wrong place.

The great thing about spending a day or two on Smith Island, besides the pleasure it brings, is the memories you take with you when you leave, which inevitably evoke the silence and isolation of the place.

Back in Crisfield, visit the **J. Millard Tawes Historical Museum,** where you'll find out that the town is famous for its oyster-shucking and crab-picking houses. These businesses get less use than in the past for the bay is depleted. Rather than being picked, many live crabs are sold to restaurants, since they bring in more money when they can be served steamed in their shells.

You can arrange for a **Port of Crisfield Escorted Walking Tour** at the museum. The town is named for John W. Crisfield, who extended the railroad into Crisfield, thus creating an explosion of activity around Tangier Sound. As a result, Crisfield gave itself the moniker "Crab Capital of the World." Crisfield is still a working waterfront town, but it has less influence than before.

Near the dock there's a little ice cream shop in **Goodsell Alley.** The alley is where prostitutes used to hang out and where many sailors got drunk and were shanghaied onto ships to work in the bay. The alley is cleaned up now, of course; only those with a vivid imagination and knowledge of the past realize that this seemingly innocent spot used to be more provocative.

At one time, Crisfield was a thriving port where steamboats docked from Virginia and Maryland's Western Shore. Now it is a reduced version of what once was. Condominiums have taken over some waterfront acreage, changing the character of the place that was once a mecca for the crab-picking and oyster-shucking industries.

Fishing charters are available from Crisfield. You can visit **Somers Cove Marina** and ask around or check in at the Somerset County Tourism Office for a list of captains. Fish caught in the waters off Maryland include rockfish, flounder, sea trout, black bass, and bluefish.

On your return to Salisbury, drive north up West Main Street (MD 413). (Going north, the road name will change from West Main Street to Richardson Avenue at 7th Street, but it is still MD 413.) After you've driven 14 miles, turn left onto US 13 for about another 20 miles.

On the way, you may want to stop at **Princess Anne,** a quiet, sleepy little college town where the **University of Maryland Eastern Shore** is located. US 13 bypasses the center of Princess Anne. In order to visit the town, take the right-hand exit for Princess Anne onto Somerset Avenue (MD 675). To reach Salisbury afterward, get back on US 13 north.

IN THE AREA

Accommodations

Alexander House Booklovers Bed & Breakfast, 30535 Linden Avenue, Princess Anne, 21853. Call 410-651-5195. Close to Crisfield in case you need a place to stay on the mainland. Three rooms with private baths. Comfortable, but there are a lot of house rules—about such things as wearing shoes inside. Web site: www.bookloversbnb.com.

Inn of Silent Music, 2955 Tylerton Road, Tylerton, 21866. Call 410-425-3541. Cheerful rooms with water views and private baths, plus delightful hosts and gourmet meals. There's no liquor license on the island, but you can bring your own wine or other alcohol. Two-night minimum on weekends. Open Apr. through Oct. Web site: www.innofsilentmusic.com.

Susan's on Smith Island Bed & Breakfast, 20759 Caleb Jones Road, Ewell, 21824. Call 410-425-2403. Overlooks the harbor with two rooms for rent. Susan is a 13th-generation Smith Islander. Open year-round. Web site: www.susansonsmithisland.com.

Attractions and Recreation

Bayside Golf Cart and Bike Rentals (with Bayside Ice Cream Stand), 4065 Smith Island Road, Ewell, 21824. Call 410-425-2771. Web site: www.smithislandcruises.com.

Boat Taxi Service, Captain Waverly Evans, P.O. Box 203, Tylerton, 21866. Call 410-968-1904. Provides transport between Tylerton and Ewell. Also sells the captain's original painted wood art pieces.

Captain Jason I, Captain Terry Laird, 4032 Smith Island Road, Ewell, 21824. Call 410-425-5931. Provides transport from Crisfield City Dock to Ewell.

The Captain Jason II *at Crisfield City Dock before trip to Tylerton*

Captain Jason II, Captain Larry Laird, P.O. Box 205, Tylerton, 21866. Call 410-425-4471. Provides transport from Crisfield City Dock to Tylerton.

J. Millard Tawes Historical Museum, 3 Ninth Street & Somers Cove Marina, P.O. Box 253, Crisfield, 21817. Call 410-968-2501. All about the seafood industry in the area. Tawes was a governor of the state who was from Crisfield. This museum is open year-round and offers wonderful walking tours of the town from Memorial Day through Labor Day. Closed Sun. Small admission charge. Web site: www.crisfieldheritage foundation.org.

Janes Island State Park, 26280 Alfred Lawson Drive, Crisfield, 21817. Call 410-968-1565 or 888-432-2267. Part of this state park is on the mainland and part is on Janes Island, with accessibility by boat only. Web site: www.dnr.state.md.us/publiclands/eastern/janesisland.html.

Smith Island Baking Company, 20926 Caleb Jones Road, Ewell, 21824. Call 410-425-2253. A commercial bakery where you can watch Smith Island cakes being created, from the layers to the fudge icing. Go early in the day if you want to watch the women baking, stacking the layers, and icing the cakes. Web site: www.smithislandbakingco.com.

Smith Island Crabmeat Co-op, Inc., 21228 Wharf Street, Tylerton, 21866. Call 410-968-1344. Open early in the morning and late in the day. Opportunity to see women crab pickers in action if crabs are in season and plentiful. Crabmeat is also for sale.

Smith Island Cruises, 4065 Smith Island Road, Ewell, 21824. Call 410-425-2771. Tour boats dock in Ewell for a few hours before making the return trip to Crisfield. Web site: www.smithislandcruises.com.

Smith Island Marina, 20880 Caleb Jones Road, Ewell, 21824. Call 410-425-4220. Many people choose to visit the island on their own boats. This leaves them free to explore on their own terms, separate from the restrictions of the ferry schedules. The marina has six slips and a small boathouse. Web site: www.smithisland.us.

Smith Island Visitor's Center & Cultural Museum, 20846 Caleb Jones Road, Ewell, 21824. Call 410-425-3351 or 800-521-9189. This small museum is actually in the community of Ewell—the largest of the three communities that make up Smith Island. With little to do when visiting the island, this museum provides both information and distraction. Open daily May through Oct. Web site: www.smithisland.org.

Somers Cove Marina, 715 Broadway, Crisfield, 21817. Call 410-968-0925 or 800-967-3473. Web site: www.dnr.state.md.us/publiclands /eastern/somerscove.html.

Dining/Drinks

Bayside Inn, 4065 Smith Island Road, P.O. Box 41, Ewell, 21824. Call 410-425-2771. Absolutely superb crabcakes made with local crabmeat. String beans and coleslaw are excellent too. For a quiet meal, time your visit in between tour boat stops unless you arrive on one of tour boats; it's much busier when the boats dock but the restaurant can handle it. Web site: www.smithislandcruises.com.

Beach to Bay Seafood, 12138 Carol Lane, Princess Anne, 21853. Call 410-651-5400 or 866-651-4505. It isn't fancy, but they have good food, including steamed hardshell crabs.

Drum Point Market, 21162 Center Street, Tylerton, 21866. Call 410-425-2108. This local grocery and hangout is the only place to grab lunch in Tylerton. Food is just adequate but local color is wonderful.

Flying Crane Café, 11747 Somerset Avenue, Princess Anne, 21853. Call 410-651-3131. Pizzas, sandwiches, and salads. Closed Sun.; hours vary. Web site: www.theflyingcranecafe.com.

Ice Cream Gallery & Gift Shop, 5 Goodsell Alley, Crisfield, 21817. Call 410-968-0809.

Peaky's Restaurant, 30361 Mt. Vernon Road, Princess Anne, 21853. Call 410-651-1950. Good food and a comfortable atmosphere. But skip the Smith Island cake; there are better versions on Smith Island itself. Web site: www.peakys.com.

Other Contacts

Crisfield Chamber of Commerce, 906 West Main Street, Crisfield, 21817. Call 410-968-2500 or 800-782-3913. Web site: www.crisfield chamber.com.

Crisfield Municipal Parking Lot, 1010 West Main Street, Crisfield, 21817. Parking is free during the day, but if you're spending the night on Smith Island, you need to buy a $5 per night parking permit from the Crisfield Police Department or Somerset County Tourism—Crisfield Office (across the street).

Crisfield Police Department, 319 West Main Street, P.O. Box 270, Crisfield, 21817. Call 410-968-1543. If it's Sunday or a holiday and the tourism office is closed, you can purchase overnight parking permits here for the municipal city lot. Web site: www.cityofcrisfield-md.gov.

Somerset County Tourism, 11440 Ocean Highway, P.O. Box 243, Princess Anne, 21853. Call 410-651-2968 or 800-521-9189. Web site: www.visitsomerset.com.

Somerset County Tourism—Crisfield Office, 1003 West Main Street, Crisfield, 21817. Call 410-968-1543. You can purchase overnight parking permits here for the municipal lot across the street.

The ever-present boardwalk in Ocean City

CHAPTER

2

Dramatic Atlantic Coastline

Traversing the Ocean City Beach, Boardwalk, and Beyond

Estimated length: 150 miles
Estimated time: Weekend

Getting there: Easy, easy, easy, for all roads appear to lead to Ocean City. From the Annapolis area, take US 50 east and cross the Chesapeake Bay Bridge. Continue across the Kent Narrows Bridge. At the fork in the road, take the right-hand route toward Easton and just stay on US 50 east.

Highlights: Watching wild ponies on Assateague Island. Playing in sand and waves on the oceanfront beach. Riding bikes on the boardwalk. T-shirt shops, french fries, and bikini-clad bathers. The charming town of Berlin, where Julia Roberts and Richard Gere shot the film *Runaway Bride*. Lots and lots of mini-golf. A canoe trip on the Pocomoke River or a kayak excursion on Ayers Creek.

With millions of travelers visiting **Ocean City,** the small beach town is hopping in warm weather. From Memorial Day to Labor Day, this beach and boardwalk locale is crowded with excited kids of all ages. However, the shoulder seasons are perhaps the best time to visit, when weather is still warm but you can walk on the beach or sit on a pier at the southern inlet without seeing more than a handful of people. If you're traveling with a dog, this is also the time when your favorite pooch is allowed to accompany you to the boardwalk (on a leash, of course).

The shoulder seasons, about mid-April through Memorial Day on one end and Labor Day through the end of November on the other, are also a great time to explore the towns beyond Ocean City. Summertime is fine too if you like crowds; on summer weekends in Maryland all roads don't lead to Rome, but instead to "OC," as Ocean City is affectionately called. Most places are open seasonally, but enough are open year-round that you can always find at least a few open restaurants and stores.

In Ocean City itself, you have the option of finding a hotel room or condo rental on either side of **Coastal Highway,** which runs the 10-mile length of the city from the inlet at the southern end to the Delaware line on the north. Oceanside rooms tend to be pricier than ones fronting Isle of Wight Bay or Assawoman Bay, though often the ocean and the bay are only a few blocks away from one another, since OC is only two or three blocks wide in some places. Like other popular oceanside resorts, the ocean and the sandy beach are the major draws, so those views come with a premium.

Wherever you stay, there's nothing quite like walking barefoot on the beach with the hot sand between your toes. It's great fun walking along the edge of the tide as waves lap at your feet and you alternately get wet and dry.

There's nothing quite like a sunrise over the ocean either, or a sunset over the bay in Ocean City with the nighttime birds squawking and flying and nesting as the orange and red orb of the sun descends into the darkening sky.

There's nothing quite like walking around with other people wearing scanty tops and shorts and bathing suits, the pervasive smells of salt water and suntan lotion, and the sight of children playing with colorful kites.

There are other beaches on the East Coast, of course—but for Marylanders and for many OC lovers from out of state, this is the place.

You'll probably see lots of references to the **Delmarva Peninsula** everywhere. That's an abbreviation for the peninsula shared by Delaware, Maryland, and Virginia, which is where Ocean City is located. The communities along the peninsula have much in common—the Atlantic coastline, related social and cultural history, and weather.

Across the bridge connecting Ocean Gateway (US 50) to Ocean City is an area known as **West Ocean City,** with a harbor where you can see large workboats—trawlers, lobster boats, and clam boats. The view is relaxing and it's a great place to unwind for a few hours away from the chaos that is the Coastal Highway corridor. West Ocean City has lots of restaurants that specialize in crab and other seafood—Sunset Grille, Captain's Galley, Harborside, the Shark, and Crab Alley—some open year-round.

Then there is the 3-mile boardwalk. You can park at the Inlet lot at the south end or in a spot behind your hotel or rental house. Bike riding is permitted on the boardwalk for a few hours in the morning and there are bike rentals all over if you don't want to be bothered bringing your own. There are also plenty of places to buy T-shirts and other beach paraphernalia, plus ice cream, fudge, fries, and other goodies.

The **Ocean City Life-Saving Museum** at the south end of the boardwalk has interesting exhibits about the history of the boardwalk, sharks' teeth, and old-fashioned bathing attire. You can learn about such innovations as the 1926 photo booth machine, where you could buy a strip of four photos for 25 cents. It revolutionized photography, for people no longer had to pay an expensive photographer to capture the moment. These machines were popular in the arcades of the 1930s. For fun, you can still find one at an arcade at the lower end of the boardwalk.

Sleeping in a car or under the boardwalk or on the beach is illegal, so you'll have to rent a hotel room, condo, or beach house. There are plenty of hotel chains and upscale boutique hotels like the **Lighthouse Club**. Rates are seasonal, so you will pay the going rate in the summer; off-season, you can generally negotiate a good deal. The boardwalk end of town tends to be more lively, the north end quieter.

There are plenty of regular golf courses in the area too, but mini-golf is decidedly a beach town phenomenon. Drive up Coastal Highway, which literally runs the entire north–south length of the city, and you'll inevitably see mini-golf courses every few blocks. You'll also find tons of restaurants that serve ribs, burgers, and seafood.

How much mini-golf can you play? Luckily, there are lots of other options.

On rainy days, there's shopping at the **Ocean City Factory Outlets** just west of the US 50 bridge to the beach. The outlet mall is crowded on rainy days, but it's a great place to spend time and hunt for bargains.

Some people are beachgoers and others are explorers. If you're a little of both and find yourself in Ocean City, you should make forays into the surrounding towns and countryside.

Once outside OC, a favorite is the nearby town of **Berlin**—just 5 miles and about 20 minutes away. Named "Hale" in *Runaway Bride* with Richard Gere and Julia Roberts, Berlin is as quaint and charming as its depiction in the film, which was shot on location in the town. At one time, the town's theme was even "Runaway to Berlin," but they've since changed it to a less memorable slogan. *Tuck Everlasting* was filmed in the town too.

To reach Berlin from Ocean City, take US 50 west to US 113 south. Turn right onto Old Ocean City Boulevard and left onto Main Street. Park and then go stand on the porch of the **Atlantic Hotel.** Townspeople walk by and greet everyone and it's almost like you've been transported to Mayberry—almost, but not quite, for Berlin, while friendly and small-townish, is up to speed on the amenities that make tourism its number one industry.

In the summer, Berlin gets visitors from OC who want a day away from the beach scene. Yet during the shoulder season, these tourists appear in dribs and drabs, so you can have the sense of discovering Berlin all for yourself. (The town is not named for the city in Germany, as many think, but rather for the Burleigh Inn, which was once prominent in town. Burleigh Inn was shortened to Burl-Inn, and the spelling was eventually changed to Berlin.)

There's a walking tour of *Runaway Bride* sites that you can do on your own; just pick up one of the brochures displayed around town. Highlights include the Atlantic Hotel, where there is a Richard Gere room, and the Rayne's Reef diner, where Julia Roberts sat at the counter trying a variety of egg dishes to determine her favorite. Keep in mind that this is a small town, so if you try to order any of the dishes, like egg whites, yourself, the waitress might say they don't do them that way. Ask for the owner next, for he promises they will serve all the egg dishes from the film, except poached eggs.

Kayaking and canoeing are prevalent in the area, and can be a great activity for a few hours. These paddling excursions are not expensive, and there are several companies that provide the boats, paddles, life preservers, and guides unless you're particularly proficient and want to go it alone.

Another side excursion of interest will take you to **Assateague Island National Seashore** where the wild ponies are reputed descendants of horses from a European boat shipwrecked centuries ago. Along with Assateague State Park, the National Seashore is located on the part of Assateague Island that belongs to Maryland; at the southern end of the island, which is part of Virginia, is the Chincoteague National Wildlife Refuge. The Native American heritage of this land is evident from the names.

One way to escape the beach crowds is to visit Assateague. Assateague has hiking and camping, but the bugs on this marshy piece of barrier island are nasty, both the mosquitoes and the tiny gnats known as "no-see-ums" that create quite a wallop on the itch scale. Cover up and spray yourself liberally with bug repellent, whatever the time of year or day. There are marshes all over and mosquitoes follow the ponies, which is what you'll be doing as well.

Unlike the bugs, the ponies will ignore you, which lets you get close enough to shoot close-up photos and to observe that they are ragged and obviously not bred like the thoroughbreds raised on horse farms throughout the state. These are wild animals, after all, and interesting for just that fact. Remember not to try and pet them or feed them—for your protection and for theirs.

Assateague is actually part National Seashore and part State Park. There's no specific delineation apparent, but it's there, and you can drive part way and then walk where you can't drive. It's helpful to stop at the visitors center if you have any questions. And if you're so inclined, you can register to camp. Otherwise, just visit for a few hours to watch the ponies ignore you.

To reach Assateague, take US 50 west from Ocean City. Turn left onto Stephen Decatur Highway (MD 611) south and cross the Verrazano Bridge over Sinepuxent Bay. From there, follow the signs.

Nearby **Snow Hill** is another quiet small town without much to see

Wild ponies seem nonchalant even with people nearby.

except for some pretty houses and a nice little independent bookstore. The town name is misleading, however, for there is no hill; it's flat like most of the Eastern Shore. From here you can go paddling on the **Pocomoke River.** To reach Snow Hill from Berlin, take US 113 south. Once in Snow Hill, you can find **Pocomoke River Canoe Company** by turning right on MD 12; the canoe company is located next to the drawbridge over the Pocomoke. **Ayers Creek Adventures** also rents canoes and kayaks close by. Whether you go paddling in a canoe or a kayak or just view the river from the shore, there are birds, turtles, and abundant fish, and the area is picturesque.

From Snow Hill, proceed 5 miles north on MD 12 and turn left onto Old Furnace Road. **Furnace Town** is 1 mile ahead on the left at the intersection with Millville Road. Since it's so close, you might as well pass through. A former company town where everyone worked for the furnace company, Furnace Town has a small church, a weaver's cottage, a blacksmith shop, and other structures; tradespeople do demonstrations for visitors, answer questions, and sell small pieces of their various arts.

CHESAPEAKE BAY BRIDGE

Regulated by the Maryland Transportation Authority, the first Bay Bridge was built and open to traffic by 1952. As traffic on the bridge greatly increased, a second bridge was completed in 1973. On most days, one bridge goes in one direction, the other in the other direction. If traffic is particularly heavy, the traffic flow can be adjusted so that one of the bridges goes in two directions, thus alleviating at least some of the tie-ups.

You pay a toll going east onto the Eastern Shore, but not going west back toward Annapolis.

Eastbound from the Annapolis area, you cross the bridge and drive onto Kent Island. A few minutes later, you cross the much smaller Kent Island Bridge, which crosses a small waterway—Kent Narrows.

Traversing the Bay Bridge is a pleasure when traffic is light; it's much less fun on summer weekends, when backups are legendary. Often the much smaller Kent Island Bridge causes the hold-up for summertime weekend traffic, as it does not have the capacity of the two spans of the Bay Bridge.

You can get to Ocean City without crossing the bridge if you go around the upper portion of the bay and come down through the Eastern Shore or through Delaware. Or you can take a boat across the bay. Or maybe you're starting from a place already on the Eastern Shore. But for many, "the bridge" means none other than the Bay Bridge and it can cut hours off travel time.

Deck at the Lighthouse Club overlooking Isle of Wight Bay

The great thing about this area of the state is that you are likely to leave wanting more . . . which will no doubt bring you back and will, at the least, provide wonderful memories of an old-style seaside sojourn, with ice cream, sand, mini-golf, and the fantastic and mercurial Atlantic Ocean. Plus all the interesting side trips you took.

To return to the Annapolis area, take US 50 west and follow the signs to the Bay Bridge. In season on the way back from the shore, you should stop at one of the many roadside stands selling local produce—white Maryland corn, watermelon, cantaloupe, peaches, tomatoes, squash, and more.

IN THE AREA

Accommodations

The Lighthouse Club at Fager's Island, 56th Street on the Bay, Ocean City, 21842. Call 410-524-5400 or 888-371-5400. A small, intimate, and upscale hotel on the Bay side of Coastal Highway built to look like—what else—a lighthouse. Web site: www.fagers.com/hotel.

The Atlantic Hotel, 2 North Main Street, Berlin, 21811. Call 410-641-3589. This old-fashioned hotel, built in 1895, has since been renovated and it is fairly modern, with wireless Internet. Yet in the bathrooms, the hot and cold water come out of different spigots. The incongruity is charming. Web site: www.atlantichotel.com.

Attractions and Recreation

Assateague Island National Seashore, 7206 National Seashore Lane, Berlin, 21811. Call 410-641-1441 or 410-641-3030. Make sure to bring your insect repellent. This barrier island has wild ponies, camping, and great photo ops. Web site: www.nps.gov/asis.

Assateague Island State Park, 7307 Stephen Decatur Highway, Berlin, 21811. Call 410-641-2120 or 888-432-2267. Open daily Memorial Day through Labor Day. Watch the birds and the ponies that live untethered on the island; the ponies get close to visitors, but don't pet them—they are wild. Web site: www.dnr.state.md.us/publiclands/eastern /assateague.html.

Ayers Creek Adventures, 8628 Grey Fox Lane, Berlin, 21811. Call 443-513-0889 or 888-602-6288. Canoe and kayak rentals for paddling on your own or with a guide on Ayers Creek in Berlin or Isle of Wight Bay off Fager's Island in Ocean City. Web site: www.ayerscreekadventures.com.

Furnace Town Living Heritage Museum, 3816 Old Furnace Road, Snow Hill, 21863. Call 410-632-2032. A glimpse into the 19th-century industrial village that operated around the Nassawango Iron Furnace—a bog ore furnace built in 1832. You can walk around and watch live demonstrations by blacksmiths, weavers, gardeners, broom makers, and woodworkers; these artisans sell some of their crafts in the gift shop on site. Open daily April through Oct. Web site: www.furnacetown.com.

Ocean City Factory Outlets, 12741 Ocean Gateway (US 50), Ocean City, 21842. Web site: www.ocfactoryoutlets.com.

Ocean City Life-Saving Station Museum, 813 Boardwalk (at the Inlet), Ocean City, 21843. Call 410-289-4991. If it's raining outside, or even if it's not, this small museum at the bottom end of the boardwalk is quite interesting. Open daily May through Oct.; weekends during other months. Web site: www.ocmuseum.org.

Pocomoke River Canoe Company, 2 River Street (at the Pocomoke River Bridge), Snow Hill, 21863. Call 410-632-3971. Canoe and kayak rentals for novice and experienced paddlers. Open daily Apr. through Oct. Web site: www.atbeach.com/amuse/md/canoe.

Dining/Drinks

Fager's Island, 201 60th Street on the Bay, Ocean City, 21842. Call 410-524-5500. Lunch and dinner. Web site: www.fagers.com.

The Globe, 12 Broad Street, Berlin, 21811. Call 410-641-0784. Formerly a theater, the Globe is a good place to go at night, with its friendly bar scene and attentive service. There's really no other nightlife in the town of Berlin, so this is the spot to be. The upstairs doubles as an art gallery for work by local artists and some nights there is live music on the stage downstairs. Lunch, dinner, and Sunday brunch. Closed Mon. Web site: www.globetheater.com.

Rayne's Reef, 10 North Main Street, Berlin, 21811. Call 410-641-2131. Counter and table seating for breakfast and lunch; soda fountain and grill food. Web site: www.raynesreef.com.

Seacrets, 117 West 49th Street, Ocean City, 21842. Call 410-524-4900. Lunch and dinner plus a popular nighttime scene with live bands and DJs. Web site: www.seacrets.com.

Other Contacts

Ocean City Department of Tourism, 4001 Coastal Highway, Ocean City, 21842. Call 410-723-8617 or 800-626-2326. Web site: www.ococean.com.

Worchester County Tourism, 104 West Market Street, Snow Hill, 21863. Call 410-632-3110 or 800-852-0335. Web site: www.visitworcester.org.

A street in Cambridge, where the houses used to look alike

CHAPTER

3

Cambridge and Salisbury

Finding Quirky Charm off US 50

Estimated length: 120 miles
Estimated time: A weekend

Getting there: From the Chesapeake Bay Bridge, it's an easy trek along US 50 east to Cambridge. Continue farther on US 50 east to reach Salisbury. In order to explore the quirky and pleasurable attractions, you'll detour off US 50 to surrounding roads and countryside.

Highlights: A fantastic waterfront Hyatt. The Salisbury Zoo. Annie Oakley's old house. The Delmarva Shorebirds minor league baseball team.

Too often people traverse US 50 east from the **Chesapeake Bay Bridge** to Ocean City, bypassing the town of **Cambridge** and the city of **Salisbury**. This trip is going to change that.

By taking detours off US 50, you see charming towns and villages, farmland and river shores. US 50 is not the end all and be all of the **Eastern Shore,** but it is a kind of lifeline and a way to get around quickly. It's important, there's no doubt of that, but there are treasures hidden off this main road too.

Plan to stay at the **Hyatt Regency Chesapeake Bay Golf Resort, Spa and Marina.** To get there, cross the Choptank River Bridge on US 50 east, and continue driving about a mile farther. Turn left into the resort on Heron Boulevard and travel up the impressive drive. When you check in, you'll do so in an unpretentious high-ceilinged lobby made of stone and wood.

As its name suggests, this is a self-contained resort—with a marina, excellent restaurants, swimming, fly-fishing, golf, and a terrific spa. Once you've checked in, resist the temptation to stay put. Instead, go out and explore.

You'll want to spend time at the Hyatt of course, for it is absolutely lovely, situated as it is along the banks of the **Choptank River.** Most of the guest rooms look out on the water. This is particularly fun during a storm, when it's fascinating to watch lightning over the river from the comfort and shelter of your room. On nights with good weather, you can sit beside the outdoor fireplace and roast marshmallows.

But again, resist the temptation to stay put at the Hyatt; it's lovely, but so are many of the attractions nearby, which you can get to by going right, straight, or left out of the entrance to the resort.

First, go right to check out the quirky town of Cambridge, where **Annie Oakley** once shot ducks from her house on Bellevue Avenue (this tale has been historically validated). While she did most of her sharpshooting out west, at one point she temporarily retired from the sharpshooting business and settled in Cambridge with her husband. But then she apparently decided to practice her shooting skills from the modest house, reportedly shooting ducks and other waterfowl from the upstairs windows (though some accounts say it was from the porch). After a few years, she left Cambridge and went back to being a sharpshooter.

To see her former house, turn right out of the Hyatt onto US 50 west for 0.7 mile. Turn left onto Washington Street for 0.9 mile, right at Race Street for 0.6 mile, continue on Locust Street for 0.6 mile, right at Glenburn Avenue 0.4 mile, left at Hambrook Boulevard 0.5 mile, and continue onto Bellevue Avenue 0.4 mile. Oakley's modest former house is privately owned, but you can stop and admire the house, which is across the street from the Choptank River and has dramatic views. One can imagine Oakley taking aim from her house at unsuspecting waterfowl.

Cambridge has lots of other old houses and great stories too. For example, the owner of the house at 1106 Locust Street painted his house with red, white, and blue stars and stripes one Fourth of July to object to the historical society's tight regulations; the flag-like paint job was quite a tourist attraction.

You never know exactly what you're going to find, for Cambridge is a town with a sense of humor, judging by at least some of the residents. Quirky is not bad, it's entertaining. There are no meters in Cambridge, so you can park on the street for free and explore at your leisure. There's an interesting decoy shop in town—**Chesapeake Classics**—if decoys (used to hunt waterfowl) are your thing. There are also a few art galleries and antiques shops. This is not a shopping mecca, but rather a sleepy little town with some unusual "stuff."

To get to Cambridge's historic downtown from the Hyatt, turn right onto US 50 west. Before the Choptank River Bridge, turn left onto Maryland Avenue and cross a small drawbridge over Cambridge Creek into downtown. Go straight through the light after the drawbridge and turn left onto Gay Street. There is free parking on the right. If you walk around for a few blocks, you'll see stores and restaurants; and you're only 5 or 10 minutes from the hotel.

After checking out the downtown area, go back to the Hyatt to swim in one of the pools or have a massage at the spa. Then when you're ready for your next excursion, leave the resort and go straight this time, crossing US 50 (Ocean Gateway) at the light and heading west on MD 16 toward Church Creek and **Blackwater National Wildlife Refuge.**

One of the first things of interest will be **Emily's Produce.** It's a terrific fruit and vegetable stand on the right-hand side just 5 miles from the Hyatt. If you're there during May, you can pick your own strawberries. If you're there in the fall, you can pick your own pumpkins. If you have young children with you, there's a small playground out back.

Go straight on MD 16 if you like cycling and kayaking and other outdoor recreation, as this is the area for you. There's good leisure cycling as well as endurance cycling via backroads that have limited traffic. The Hyatt concierge can help direct you.

Continue driving past the produce stand toward **Taylor's Island.** When you've traveled 16 miles from the entrance to the Hyatt on MD 16, you'll come to **Slaughter Creek Marina.** The name of the marina may sound ominous, but if you turn right into the marina, there's a floating restaurant that used to be a floating Coast Guard station in Florida. The restaurant is **Palm Beach Willie's** and it offers both indoor and outdoor seating, plus good food. When the weather is rough, you can feel the building sway, but that is part of its charm.

On the way to the restaurant, you will have driven by and through

Some local goodies for sale at Emily's Produce

land that is part of Blackwater National Wildlife Refuge. If you're interested, there are trails and opportunities for biking. The goal of the refuge is research and habitat restoration, but there are some trails for recreational use. Birders will definitely find plenty to watch in the wildlife refuge, where there are fields, woods, and swamps.

After a relaxing meal at Palm Beach Willie's, you're only 16 miles from the Hyatt. So you can head back to the resort for some relaxation. Play a game with the kids. Have a drink at the bar. Sit and look out at the river. Work on your tan.

When you're ready to go out again, next time you'll go left…to **Salisbury.**

Salisbury is about 30 miles west of Ocean City and about 70 miles east of the Bay Bridge, so to reach it, head east on US 50. Salisbury is also 10 miles from the Delaware line along US 13. Thus it is at a crossroads…and has become the shopping mecca of the Eastern Shore, for it has the shore's only major mall (except for the outlet malls outside Ocean City and near the Bay Bridge).

In addition to shopping, Salisbury has the largest year-round population on the Eastern Shore. The city is located at the head of the Wicomico River, which leads to the Chesapeake Bay. And there's a large university there—Salisbury University, with a really pretty campus.

In many ways, Salisbury is a hub of the Eastern Shore. In additional to the mall (**Centre at Salisbury**), there's not just one but two WalMarts. Many people from rural areas throughout the Eastern Shore travel to Salisbury to go to one of the WalMarts. If they have time, they'll go to both WalMarts. Because of the shopping, it is the "Big City" on the Eastern Shore. By big-city standards, Salisbury is little, but for the Eastern Shore, it represents a center of commerce, where everyone goes to shop, or to a doctor, or for other business and personal needs.

It's not unusual to hear people exclaim, "They've got two WalMarts!" This is important to all the surrounding communities. And then there's nearby Ocean City, which gets extremely crowded, especially during the summer. While that can be fun, there are nearby places where you'll have more breathing room, like Salisbury.

From the Hyatt in Cambridge, Salisbury is an easy drive if you're looking for some fun. There's no ocean, but there is a river, great food, and lots to do.

There's the **Salisbury Zoo,** for instance. It's a wonderful regional zoo—a must-see and must-do. The zoo is charming and larger than one would expect. It's well done. The animals are interesting and look, for the most part, as if they are well cared for. The zoo is free to visitors and it's in a park, so you can bring a picnic lunch and let the animals' antics entertain the kids. Or go yourselves if there are no kids along; animals are fun for adults too, especially in a good zoo.

US 50 (east–west) and US 13 (north–south) intersect at Salisbury and are lifelines to the entire area. Right near the intersection of these two major highways, Salisbury has **Arthur W. Perdue Stadium,** home to one of the state's minor league baseball teams, the **Delmarva Shorebirds.** During the seventh-inning stretch at Shorebirds home games, everyone stands to stretch, as is usual, and then does "the chicken dance," which is not so usual. The chicken dance is a nod to the team's name and to the many chicken farms nearby.

It feels good to sit outside on a summer evening without the larger crowds of a major league game, but in a decent-sized stadium nonetheless …as it cools off from the hot sun of the day. Of course, planning to attend a game has the inherent risk of being rained out, so have a backup plan.

But if the weather cooperates, the local crowd, plus comfort food like pizza, hot dogs, peanuts, and ice cream, might be just the ticket. The team feels like an Orioles farm team, which it is. Perhaps it's the black and bright orange colors, which mimic those of the Orioles.

Cool breezes, lights on the scoreboard, kids chasing foul balls. This is as good as it gets, as Americana, as wholesome. Two umpires—one behind home plate and the other in front of the shortstop—call the game. Young

Minor league baseball at Delmarva Shorebirds home stadium in Salisbury

boys and girls proudly sport their own baseball gloves brought from home. There's a bird mascot—what else? Fans bring their enthusiasm and appetite for a good time. There's a baseball museum and shop inside. Almost everyone does the chicken dance in the middle of the seventh inning, except for those from outside the area who are merely visiting and may not have the moves down. It's funny to watch. This is a spectator sport after all.

Perdue Stadium is the perfect spot to watch the minor league Delmarva Shorebirds play baseball. The stadium is located across the street from the Perdue Corporation (of chicken fame). With so many chicken farms in this part of the state, Frank Perdue made his fortune processing them.

At the stadium, the food concessions are not great, but the ballgames are fun. Plan on eating somewhere else before or after the game; you'll have a better meal. And you have to pay for parking in the lot at the stadium, but it's only a few dollars.

When you're on the Eastern Shore near the Atlantic Ocean and you ask residents of small towns where they go when they want to visit a city—the answer is Salisbury, which is, in reality, rather small. "But it has everything we need" is often the answer. How can you argue with that?

Market Street Inn is an excellent restaurant you might want to try before you leave Salisbury. Get on the business portion of US 50 east. At the intersection with Division Street turn right. Go through the four-way stoplight and turn right at the next light onto Circle Avenue. The restaurant is on the left at the corner of Carroll and Market Streets.

To go back to the Hyatt, get back on US 50 going west and follow it to the light at Heron Boulevard, where you turn right onto the resort property. Now that you've explored everywhere else, wander around the extensive Hyatt property and make use of all the facilities. This trip was meant to merely enhance the Hyatt, to demonstrate that there's more in Cambridge than the one resort, even though it's a fantastic resort. That's as it should be …to have more great choices than time to do them all. To want to return, and to devise plans for the next visit…before this one is even done.

On the way back to the Annapolis area, there's a large outlet mall in **Queenstown** right before you get back on the Bay Bridge. If you're traveling at rush hour and you want to wait for traffic to clear, or if it's shopping you want, you can stop and shop at the outlets on the way home after your weekend.

IN THE AREA

Accommodations

Hyatt Regency Chesapeake Bay Golf Resort, Spa & Marina, 100
Heron Boulevard, Cambridge, 21613. Call 410-901-1234 or 800-233-
1234. Web site: www.chesapeakebay.hyatt.com.

Attractions and Recreation

Annie Oakley's House, 28 Bellevue Avenue on Hambrooks Bay, Cambridge, 21613. Privately owned but you can drive by. This 1912 house was
designed and constructed by Wild West sharpshooter Annie Oakley. She
lived here with her husband, Frank Butler, for several years.

Arthur W. Perdue Stadium, 6400 Hobbs Road, Salisbury, 21802. Call
410-219-3112. Home field for the Delmarva Shorebirds, a Class A minor
league affiliate of the Baltimore Orioles. Web site: www.theshorebirds
.com.

Blackwater National Wildlife Refuge, Visitors Center, 2145 Key Wal-
lace Drive, Cambridge, 21613. Call 410-228-2677. More than 27,000 acres
of forest, marsh, and freshwater ponds. A haven for birds. Web site: www
.fws.gov/blackwater.

The Centre at Salisbury, 2300 North Salisbury Boulevard, Salisbury,
21801. Call 410-548-1600. Major shopping mall for the Eastern Shore
with department stores, clothing, electronics, restaurants, and more.
Located on US 13 north of the bypass; accessible from both US 13 and US
50. Open daily. Web site: www.centreatsalisbury.com.

Chesapeake Classics, LLC, 317 High Street, Cambridge, 21613. Call
410-228-6509. Waterfowl decoys, fish decoys, and lures. Web site: www
.chesapeake-classics.com.

Dorchester County Courthouse, 206 High Street, Cambridge, 21613.
Call 410-228-1000. Active courthouse on the site where numerous slave
auctions were held during the antebellum period. Web site: www.tour
dorchester.org.

House in Cambridge once owned by Annie Oakley and her husband

Eastern Shore Baseball Hall of Fame Museum, at the Shorebirds' Stadium, 6400 Hobbs Road, Salisbury, 21802. Call 410-546-4444.

Prime Outlets-Queenstown, 441 Outlet Center Drive, Queenstown, 21658. Call 410-827-8699. Superb and extensive outlet mall close to the Chesapeake Bay Bridge off US 50. Open daily. Web site: www.prime outlets.com/queenstown.

Salisbury University, 1101 Camden Avenue, Salisbury, 21801. Call 410-543-6030. Originally the Normal School and then Salisbury State Teachers College, this university has grown in size and stature. It has a pretty campus for both undergraduate- and graduate-level students. Web site: www.salisbury.edu.

Salisbury Zoological Park, 755 South Park Drive (in City Park), Salisbury, 21802. Call 410-548-3188. This is a charming must-see, especially if you have young children. Even adults enjoy walking through this lovely regional zoo. Mammals, birds, and reptiles. Open daily except Thanksgiving and Christmas. Free admission. Web site: www.salisburyzoo.org.

WalMart Supercenter North, 2702 North Salisbury Boulevard, Salisbury, 21804. Call 410-860-5095. Web site: www.walmart.com.

WalMart Supercenter South, 409 North Fruitland Boulevard, Fruitland, 21801. Call 410-341-4803. Web site: www.walmart.com.

The Ward Museum of Wildfowl Art, 909 South Shumaker Drive, Salisbury, 21804. Call 410-742-4988. Dedicated to the artistry of sculpted decoys used by hunters to attract and entice migrating ducks and geese. The museum is part of Salisbury University. From US 50, follow Beaglin Park Drive 1.2 miles. Open daily. Web site: www.wardmuseum.org.

Dining/Drinks

Emily's Produce, 2206 Church Creek Road (MD 16), Cambridge, 21613. Call 410-228-3512 or 443-521-0789. Fresh fruit and vegetables. Free children's play area. Open May through Oct. Web site: www.emilysproduce .com.

Market Street Inn, 130 West Market Street, Salisbury, 21801. Call 410-742-4145. Good food and a pleasant setting on the Wicomico River. You can sit inside, outside, or at the bar. Jambalaya is good. Easy to find, from Cambridge via US 50 east–business route. On-site parking. Open daily for lunch and dinner. Web site: www.marketstreetinnsalisbury.com.

Palm Beach Willie's, 638 Taylors Island Road (at the Slaughter Creek Marina), Taylors Island, 21669. Call 410-221-5111. Built on a barge that was once used by the Coast Guard; now it's a floating restaurant and bar. Closed Mon. and Tues. Web site: www.palmbeachwillies.com.

The Red Roost, 2670 Clara Road, Whitehaven, 21856. Call 410-546-5443. A former chicken coop turned crab house, the Red Roost is a great place to enjoy all-you-can-eat steamed hardshells in an isolated yet cheerful spot. Bring lots of people because crab feasts are best enjoyed in large social groups. If crabs are out of season, try a full or half slab of ribs. Dinner only; closed in the winter. Web site: www.theredroost.com.

Suicide Bridge Restaurant, 6304 Suicide Bridge Road, Hurlock, 21643. Call 410-943-4689. It's got good branding, but the food is not the greatest. Still, you come here to hear the legend and see the bridge where several locals have taken their own lives. There's a tiny marina, lots of ducks, and a large, lively restaurant. Perhaps the best bet is the all-you-can-eat steamed hardshell crabs. Don't fill up on the other stuff—stick to the crabs. It's a casual place—shorts, jeans, and polo shirts are the dress code. Web site: www.suicidebridge.com.

Other Contacts

Dorchester County Tourism Department, 2 Rose Hill Place, Cambridge, 21613. Call 410-228-1000 or 800-552-8687. Web site: www.tour dorchester.org.

Queen Anne's County Tourism, 425 Piney Narrows Road, Chester, 21619. Call 410-604-2100. Web site: www.discoverqueenannes.com.

Wicomico County Convention & Visitors Bureau, 8480 Ocean Highway, Delmar, 21875. Call 410-548-4914 or 800-332-8687. Web site: www .wicomicotourism.org.

Statue of a Confederate soldier in front of the courthouse in Easton

CHAPTER

4

Seductive Waters of the Chesapeake Bay

Crossing to Easton, St. Michaels, Tilghman Island, and Oxford

Estimated length: 160 miles
Estimated time: A long weekend

Getting there: From Annapolis, head east on US 301/50 to the Chesapeake Bay Bridge. Pay the toll and cross the 4.3-mile span to the Eastern Shore. Follow US 50 east. About 9 miles after the bridge, the road splits into 301 and 50. Bear right and continue on US 50 east (Ocean Gateway) toward Ocean City. After a while, make a slight right turn onto Dover Road, which becomes Dover Street, and follow it into downtown Easton.

Highlights: Town of Easton for dining and shopping. Tilghman Island, a watermen's community with appealing B&Bs. Town of Oxford via ferry. St. Michaels and the Chesapeake Bay Maritime Museum. Great restaurants. Beautiful waterfront scenery.

Seductive is an accurate description of this area around the **Chesapeake Bay,** with its entrenched nautical culture. Recreation is especially water-centric, with boating, fishing, crabbing, sunbathing, and dining on seafood. With so many visitors attracted to the towns on this trip, more and more shopping has turned up too.

The Chesapeake Bay Bridge is to Maryland what Golden Gate Bridge is to San Francisco—iconic, vital, and ever present. Heavily traversed, the Chesapeake Bay Bridge can get particularly crowded on summer weekends with the regular beachgoing crowds heading to Ocean City and other Atlantic beaches in nearby Delaware.

But once over the bridge, you need not go far to find interesting destinations. Of course, you can head to the Atlantic Ocean and the beach in Ocean City. But as an appealing alternative, you can look toward riverside and bayside towns for inspiration.

There is no part of the state that is any more beautiful, more charming, more geared to a getaway than the places on this trip. Each town is worthy of a weekend or more on its own, but we'll see a little of each one this time. No doubt you will want to return.

We're starting in **Easton**—a pretty, entertaining little town. You pass right by it on US 50 east, which leads from the Bay Bridge to Ocean City, the route taken by hundreds of thousands each summer week. However, not many detour off the main drag and into Easton, which is a shame, for it's such a pleasant place for a sojourn. Tons of restaurants and shops, plus several places to stay, make this a potential destination in and of itself, not just an add-on, though it can be that too.

Easton is thus a well-kept secret. Surrounding Easton is agricultural farmland, where roadside stands in season will sell corn, peaches, watermelon, cantaloupe, tomatoes, and pumpkins; sometimes these and other products are even "pick-your-own."

In-town Easton is anything but agricultural. Consider staying at the **Tidewater Inn,** which is a lovely historic hotel with flat-screen TVs and Internet connections. The Tidewater's location is perfect, for within walking distance are assorted boutique shops, restaurants, and art galleries.

Besides the places you'd expect, there are surprises. In the back of **Hill's Pharmacy,** for example, is a soda fountain with excellent ice cream, old-fashioned milkshakes, and low prices. Also within walking distance is a statue to Confederate soldiers in front of the town's courthouse. This was, after all, a conflicted state during the Civil War. And there's a gun shop where former vice president Dick Cheney is known to buy ammunition for hunting; he owns a house in nearby St. Michaels.

On the restaurant scene, make sure to have dinner at **Out of the Fire.** The paella is superb but there are many other great dishes in this restaurant, which is so unlike most of the crab houses and other down-home restaurants on the Eastern Shore. Out of the Fire has a kitchen bar, a wine

Hooper Strait Lighthouse at the Chesapeake Bay Maritime Museum in St. Michaels

bar, and traditional tables, giving patrons options for seating, along with interesting options for food, including pizzas; the restaurant has the feel of a dining spot in a large cosmopolitan city, but it's in this little town.

After spending a day and perhaps a night in Easton, head to **St. Michaels.** From US 50, exit onto MD 322 (Easton bypass). Follow to exit for MD 33 to St. Michaels.

St. Michaels, on the **Miles River,** is only about 15 minutes away from Easton, but it's another type of place, since it is a large boating and waterfront community. There's a great crab house—the **Crab Claw**—that you should try. While it looks touristy, lots of locals go to the Crab Claw and the food is excellent. You can let the servers cover your table with white paper and bring you steamed hardshell crabs and beer, or leave the table as is if you want to order more upscale fare like rockfish topped with crabmeat. Indulge in an ear of white Maryland corn if it's in season; you'll likely want seconds.

Right next door to the Crab Claw is the large **Chesapeake Bay Maritime Museum.** There are ten exhibit buildings and you can go at your own pace, spending as much or as little time as you like. Don't miss the 1879 Hooper Strait screwpile lighthouse, which you can climb and tour; just watch out on the top level if you have children with you—the railing is a bit shaky.

Besides eating crabs, watching boats, and visiting the museum, you'll want to soak up the atmosphere of St. Michaels by wandering around town to see the shops. Just being in St. Michaels is fun, for everyone else is wandering around too; enjoyment of such a charming place is contagious.

Next on the itinerary is **Tilghman Island,** which is reached by continuing approximately 12 miles farther on MD 33 until you cross the **Knapps Narrows Bridge**—a little drawbridge on the road. Once across the drawbridge, there's no way of getting lost, except in the peacefulness of the place, for Tilghman Island Road runs north and south. The Chesapeake Bay is on the west and the **Choptank River** and **Harris Creek** are on the east, with little land in between.

Tilghman is surrounded by water and much of the scenery is breathtaking. Historically, steamboat service reached the island in 1893, and islanders had a successful seafood-processing industry. During World War II, Tilghman Packing Company provided canned rations for the troops overseas by working round-the-clock shifts. That business is gone now. But there is still a working harbor and a recreational marina. You can go out for a lighthouse boat tour with Captain Mike, visit the nautical bookstore, dine in several wonderful restaurants, and talk to the real watermen who live and work on Tilghman. The place may inspire you, as well, to simplify your life if only for a few days. For Tilghman is truly an authentic experience.

If at all possible, plan a stay at the **Lazyjack Inn Bed & Breakfast** on Dogwood Harbor with innkeepers Mike and Carol Richards. You'll immediately feel right at home there. Mike is captain of the boat that does the lighthouse tours and he can tell you just about anything you want to know about the area. Carol is warm and gracious, and her breakfasts are terrific. This is the best kind of B&B imaginable—a beautiful place with great homey comfort. It's not surprising that many of their guests return.

The last town on this trip is **Oxford** . . . one of Maryland's oldest towns.

Once upon a time, Oxford was Maryland's largest and first official port of entry on the Eastern Shore. A prominent Oxford merchant, Robert Morris, was father of a signer of the Declaration of Independence.

But the growth of Baltimore as a major shipping port caused Oxford's success to wane. While the financial repercussions of the decline had an impact, the town, with its elegant historic homes along the **Tred Avon River,** is picture perfect. This is especially true when you arrive by ferry from Bellevue across the river.

A haven for boaters, weekend visitors, and summer residents, Oxford is now a relaxed, sleepy little town—an appealing place with beautiful water views overlooking the Tred Avon River, and close to the Choptank River. The town was a movie location for the film *Failure to Launch* with Sarah Jessica Parker and Matthew McConaughey.

You can reach Oxford from Easton by taking US 50 to Easton Parkway (MD 322), then to MD 333 south for 11 miles.

Way more fun and much more interesting is to get to Oxford from Tilghman Island. To do this, drive from Tilghman back through St. Michaels, take MD 33 east to MD 329 south and take the **Oxford-Bellevue Ferry.** The ferry claims to be the oldest privately operated ferry in the country; that's entirely possible because the ferry started in 1683. It was discontinued after the Revolutionary War but resumed operation in 1836. In warm weather, the ferry crosses the Tred Avon River every 20 minutes from sunrise to sunset. Cars, bikes, and pedestrians are taken on the delightful crossing, and the fares are modest.

Whether you're staying over in Easton, Tilghman Island, St. Michaels, or even Oxford itself, having dinner at the **Robert Morris Inn** is an absolute must. Chef Mark Salter left his 17-year stint at the **Inn at Perry Cabin** in St. Michaels to be his own boss at the Robert Morris Inn. His food is superb, a gastronomic delight. What's nice too is that the prices are not out of the ballpark. Just the food is out of sight. Crab spring rolls and cream of crab soup are decadent, as are many other dishes.

The Robert Morris Inn is located in its original 1710 structure. There are some upstairs rooms where you can stay, but they are small and outdated with tiny bathrooms. Instead there are two cottages next door that are roomier and come equipped with TVs. You may want to skip staying at the inn altogether, but don't skip the food. There's a formal dining room and a larger, more casual tavern adjacent to it; both feature the exquisite food of Chef Salter.

One other food-related fact of importance in Oxford is the presence of the most fantastic homemade ice cream imaginable at the **Scottish Highland Creamery.** The ice cream is unbelievably delicious, as are the homemade sorbets.

To finish up the trip, leave Oxford by heading away from the river on Morris Street (which becomes MD 333). Follow the road until you reach a light at the intersection with MD 322. Make a left onto 322 and follow it until it ends at US 50. Cross the eastbound lanes of US 50 and make a left onto US 50 west. Follow it across the Kent Narrows Bridge to the Bay Bridge and onto US 50/301 back to Annapolis.

IN THE AREA

Accommodations

Bartlett Pear Inn, 28 South Harrison Street, Easton, 21601. Call 410-770-3300. Comfortable place to stay and/or to dine. Web site: www .bartlettpearinn.com.

Chesapeake Wood Duck Inn Bed & Breakfast, 21490 Gibsontown Road, Tilghman Island, 21671. Call 410-886-2070 or 800-956-2070. Six guest rooms with private baths. Web site: www.woodduckinn.com.

Historic Tidewater Inn, 101 East Dover Street, Easton, 21601. Call 410-822-1300. A comfortable, historic hotel perfectly situated in the heart of downtown Easton. Web site: www.tidewaterinn.com.

Inn at Perry Cabin, 308 Watkins Lane, St. Michaels, 21663. Call 410-745-2200 or 800-722-2949. Web site: www.perrycabin.com.

Knapps Narrows Marina & Inn, 6176 Tilghman Island Road, Tilghman, 21671. Call 410-886-2720 or 800-322-5181. Clean and cheerful with views of Knapps Narrows and the Chesapeake Bay. Stay includes continental breakfast. Web site: www.knappsnarrowsmarina.com.

Lazyjack Inn Bed & Breakfast (on Dogwood Harbor), 5907 Tilghman Island Road, Tilghman, 21671. Call 410-886-2215 or 800-690-5080. Four lovely rooms with private baths offer varied views and sitting areas, making this a delightful place for a getaway year-round. Excellent breakfasts. Web site: www.lazyjackinn.com.

Ruffled Duck Bed & Breakfast, 110 North Morris Street, P.O. Box 658, Oxford, 21654. Call 410-226-5496. This cheerful B&B has friendly proprietors and three rooms for rent upstairs. The curved stairs are small and you need to watch your head on the low ceiling, but the Savannah room upstairs is a winner. Web site: www.ruffledduckinn.com.

Sandaway Waterfront Lodging, 103 West Strand Road, Oxford, 21654. Call 888-726-3292. Some rooms are better than others; ask for a large bathroom and a screened-in porch. Web site: www.sandaway.com.

The Tilghman Island Inn, 21384 Coopertown Road, Tilghman Island, 21671. Call 410-886-2141 or 800-866-2141. Beautiful setting for this 20-room inn right on the Bay. Web site: www.tilghmanislandinn.com.

The Chesapeake Bay at the southern tip of Tilghman Island

Attractions and Recreation

Academy Art Museum, 106 South Street, Easton, 21601. Call 410-822-2787. Web site: www.academyartmuseum.org.

Albright's Gun Shop, 36 East Dover Street, Easton, 21601. Call 410-820-8811 or 800-474-5502. Web site: www.albrightsgunshop.com.

Chesapeake Bay Maritime Museum, Navy Point, 213 North Talbot Street, P.O. Box 636, St. Michaels, 21663. Call 410-745-2916. This complex is a great place to become acquainted with various aspects of Chesapeake Bay history. Open year-round but days and hours vary seasonally. Web site: www.cbmm.org.

Chesapeake Lights: Lighthouse Tours, Tilghman Island, 21671. Call 410-886-2215 or 800-690-5080. Different length tours with Captain Mike Richards on his Coast Guard–approved boat. Provides a wonderful chance to see lighthouses up close from the water. Web site: www .chesapeakelights.com.

The Clay Bakers, 1 South Washington Street, Easton, 21601. Call 410-770-9091. Easton really doesn't have much for children; but if you have kids with you or want to express your own creativity, this paint-it-yourself pottery spot is cheerful and just the ticket for a few hours' entertainment. Open daily. Web site: www.theclaybakers.com.

Crawfords Nautical Books, 5782 Tilghman Island Road, Tilghman, 21671. Call 410-886-2230. Open weekends only, Apr. through Dec. Web site: www.crawfordsnautical.com.

Lady Patty Sail Charters, P.O. Box 112, Tilghman Island, 21671. Call 888-250-3030. Sails from Knapps Narrows Marina. Captain Jeff Mathias, Coast Guard certified. Web site: www.ladypatty.com.

Mystery Loves Company Booksellers, 202 South Morris Street, Oxford, 21654. Call 410-226-0010 or 800-538-0042. A great little bookstore located in a former bank building. The owners have a terrific selection for such a small store; you're bound to find something of interest. The business used to be in Baltimore's Fells Point, but the owners relocated and moved their store with them. Now the bookstore carries much more than mysteries. Closed Tues. and Thurs. Web site: www.mystery lovescompany.com.

Oxford-Bellevue Ferry, 27456 Oxford Road, Oxford, 21654. Call 410-745-9023. Crosses the Tred Avon River. Open daily from early Apr. through late Nov. Fares vary based on one way or round trip, and car, bicycle, or walk-on pedestrian. Web site: www.oxfordferry.com.

Phillips Wharf Environmental Center, 21604 Chicken Point Road, Tilghman, 21671, 410-886-9200. Terrapin turtles, horseshoe crabs, and more. Web site: www.pwec.org.

Talbot County Courthouse, 11 North Washington Street, Easton, 21601. Call 410-770-8010. Abolitionist and anti-slavery speaker Frederick Douglass gave his "Self-Made Men" speech here one night in November 1878. Douglass was born a slave nearby, taught himself to read, and obtained his freedom before working to help others do the same through the Underground Railroad. Outside the courthouse is a statue dedicated to Confederate soldiers.

Tilghman Watermen's Museum, P.O. Box 344, Tilghman 21671. Small museum that preserves and shares memories of Tilghman watermen and their families. Free admission. Web site: www.tilghmanmuseum.org.

Dining/Drinks

Crab Claw, 304 Mill Street at Navy Point, St. Michaels, 21663. Call 410-745-2900. Perfect setting for devouring tons of steamed hardshell crabs, either inside or outside along the Miles River. Open daily Mar. through Nov. Web site: www.thecrabclaw.com.

Grassini's Bridge Restaurant, 6136 Tilghman Road, Tilghman, 21671. Call 410-886-2929. Great views; children will particularly love watching the drawbridge open and close. Local paintings, photographs, and pottery are for sale. There are several small, intimate dining rooms plus a bar with a pool table. Open for lunch and dinner; closed Wed. Web site: www .grassinisbridgerestaurant.com.

Out of the Fire, 22 Goldsborough Street, Easton, 21601. Call 410-770-4777. Excellent restaurant—not to be missed. Great food and service. Web site: www.outofthefire.com.

Schooner's on the Creek, 314 Tilghman Street, Oxford, 21654. Call 410-226-0160. Good place to enjoy a beer and steamed hardshells, either sitting inside or outside on the patio overlooking Town Creek. Pinball and pool in the bar. Closed Wed.

Scossa Restaurant & Lounge, 8 North Washington Street, Easton, 21601. Call 410-822-2202. Dine inside or at the sidewalk café. Located opposite the courthouse. Web site: www.scossarestaurant.com.

Scottish Highland Creamery, 314 Tilghman Street, Oxford, 21654. Call 410-924-6298. "There are no bad flavors." Fantastic homemade ice cream and sorbets. The blackberry sorbet is particularly to die for. The sign on the building reads SCHOONER'S, the name of the restaurant next door. The ice cream is sold on the left-hand side of the building. Look for the line of people. Closed Jan. through Mar. Web site: www.scottishhighland creamery.com.

Soda Fountain at Hill's Pharmacy, 32 East Dover Street, Easton, 21601. Call 410-822-9751.

Robert Morris Inn, 314 North Morris Street, Oxford, 21654. Call 888-823-4012. With Chef Mark Salter and his exquisite food, dining here is an event. The inn overlooks the harbor and the ferry. St. Michaels is only 7 miles away; Cambridge is just 14 miles. Web site: www.robertmorrisinn .com.

Two If By Sea, 5776 Tilghman Island Road, Tilghman 21671. Call 410-886-2447. Soda fountain, dining room (bring your own booze), and antiques for sale make this a fun place to eat. The food is good too; soft shell crab piccata and barbecue ribs are two examples. Web site: www .twoifbyseacafe.com.

Other Contacts

Easton Main Street, 11 South Harrison Street, Easton, 21601. Call 410-820-8822. Web site: www.eastonmainstreet.com.

Oxford Business Association, P.O. Box 544, Oxford, 21654. Call 410-745-9023 (Oxford-Bellevue Ferry). Web site: www.portofoxford.com.

St. Michaels Maryland Business Association, P.O. Box 1221, St. Michaels, 21663. Call 800-808-7622. Web site: www.stmichaelsmd.org.

Talbot County Office of Tourism, 11 South Harrison Street, Easton, 21601. Call 410-770-8000. Web site: www.tourtalbot.org.

Tilghman Island. Web site: www.tilghmanisland.com.

The Point Lookout Lighthouse on the tip of the peninsula

CHAPTER

5

Point Lookout and St. George Island

Touring Lighthouses and Military Strongholds

Estimated length: 120 miles
Estimated time: Overnight

Getting there: From I-495 (Capital Beltway around Washington, DC) take MD 301 to MD 5 in Waldorf. Stay on MD 5 by taking a right turn south of Mechanicsville. Travel through Leonardtown outskirts and up the hill. Turn right at the light onto Washington Street (MD 245). At the end of the small, quaint town square, turn left onto Park Avenue. There's a parking lot on the left-hand side opposite the Executive Inn & Suites Park Avenue.

Highlights: Sotterley Plantation with gorgeous views of the Patuxent River. Leonardtown, a lovely small town. Point Lookout, where the Potomac River and Chesapeake Bay meet. Piney Point Lighthouse Museum. Little-known St. George Island.

This trip is an unusual one because you're going to a part of the state that remains little known. Yet there is much to recommend it. This area is both beautiful and intriguing, and the trip will be memorable.

A major source of employment in this part of the state is the Naval Air Station Patuxent River. Besides the air station's large military community, there is a large defense contractor community, many of whom do a reverse commute from the DC metropolitan area. Military activity has long been a factor in the region due to the water routes all around and the nearness of the nation's capital.

The first stop is **Leonardtown,** where there are a few interesting shops, including a used bookstore, flower shop, coffee shop, and quilt and fabric shop, plus restaurants and an old-fashioned bank building. Of particular note is a tiny hole-in-the-wall restaurant with superb food and friendly service—**Kevin Thompson's Corner Kafe.** Don't miss Kevin's Maryland crab soup and fried rockfish, or his squash casserole. You'll leave with your stomach full and your spirit revived.

After lunch, get back in the car and head to **Point Lookout State Park.** To get to the park, return to MD 5 and continue south for approximately 26 miles. On the way, you'll pass **Buzzy's Country Store,** a great place to stop. The local men hang out at Buzzy's and you'll get a feel for this rural community by walking in. You can also buy fishing gear, beer, or soda, and it's a good place for getting directions.

By the way, there are few gasoline stations in this neck of the woods, so be sure to fill your tank whenever possible.

Buzzy's Country Store, where the locals hang out

Point Lookout is at the southern tip of the Western Shore of Maryland, with the Potomac River to the west and the Chesapeake Bay to the east. When you get close to the tip, the road is on an extremely narrow piece of land with water on both sides. People go to the state park to fish, camp, swim, and picnic. There's a lighthouse too, but it's behind a locked fence as it was formerly part of a military installation. During the Civil War, this was a strategic location, with Virginia just across the Potomac River and many Southern sympathizers in this part of the state.

Also during the Civil War, the Union built a prisoner-of-war camp at Point Lookout for Confederate prisoners, many of whom died from the deplorable conditions. Two Confederate Civil War monuments, one marking the burial site of soldiers who died at Point Lookout, stand sentinel just before you enter the park. Consider paying your respects, whatever your political leanings.

FACTS ABOUT THE CHESAPEAKE BAY

- 30 miles at its widest point.
- North America's largest estuary (an arm of the ocean at the lower end of a river or group of rivers). An estuary is a mixture of fresh and salt water, creating a unique environment.
- Rather shallow—on average only 21 feet deep.

From Point Lookout, get back on MD 5 north, backtracking through St. Mary's City to the light at the intersection with MD 249. Turn left onto MD 249 and travel approximately 12 miles. Toward the end, you'll cross a small bridge onto **St. George Island.** The **Island Inn & Suites** is around the curve on your left.

First settled in 1634, St. George Island is an obscure place, but delightful and extremely dramatic with the Potomac River on one side of the road and the St. George Creek on the other. The British tried to invade the island during both the Revolutionary War and the War of 1812. Now it is a peaceful place with some full-time residents, watermen, and vacation homes. It is a tiny community and going there is relaxing. Plan to stay overnight at the small inn; the suites are comfortable and the service friendly. Connected with the inn and right next door is the **Island Bar & Grill,** where the seafood is quite good and the scene is upbeat with live music in the bar.

After dinner and a good night's sleep, check out of the inn. But before you turn right to head off the island, turn left instead and drive around a bit. The island homes are rather interesting and the location idyllic, though during a storm one can easily imagine the low-lying land being overrun by the forces of water.

After you've seen the island, turn around and head back on MD 249 north across the bridge. When you reach Lighthouse Road, make a left and follow the road to the end at **Piney Point.** You will pass more interesting waterfront homes through this stretch, and here too, one can easily imagine the devastation that can come from a storm. But the views are spectacular.

At the **Piney Point Lighthouse, Museum and Historic Park,** you can tour the grounds, see the lighthouse, look for osprey nests out on the Potomac River, and wander around the small museum and boat collection. It's really interesting, whether or not you are a lighthouse aficionado. The remoteness is particularly moving.

After becoming acquainted with lighthouse lore, head back the way you came on MD 249 going north this time. When you reach MD 5, turn left (north) and travel approximately 12 miles back to Leonardtown. Turn right at the light onto MD 245 north. Follow the road through the intersection with MD 235 and continue on MD 245 about 3 miles to **Sotterley Plantation** on your right.

Sotterley Plantation is terrific. Make sure to take the house tour; it is well worth it, for the volunteer docents have a wealth of information. The house and grounds have gorgeous views of the Patuxent River and it is easy to imagine the joy at waking up to such views. There are beautiful old trees around the house too. What a beautiful site, yet all is not exactly as it seems, for slavery was a fact of life here back before the Civil War. There's still one slave cabin left standing on the property. It's unusual to find an intact slave cabin, so you should walk over and take a look. It's not far from the main house.

Anyone who has been to Thomas Jefferson's Monticello in Virginia will have a sense of the scope and grand vistas you can expect at Sotterley Plantation, which is also on a large plot of land and has many gardens (and had slaves who worked there). Sotterley is certainly less well known than Monticello and a U.S. president never lived there; still it is valuable both historically and culturally.

By now, you're probably hungry. Proceed back up MD 245 and turn left onto Three Notch Road. Turn left again onto Mervell Dean Road and left once more onto Clarkes Landing Road. At the end of the road, turn left into the restaurant parking for **Clarke's Landing Restaurant** in the town of

The main house at Sotterley Plantation along the Patuxent River

Hollywood. Seemingly in the middle of nowhere, this seafood restaurant and crab house is off the Patuxent River. There's a dock where boaters can pull up to get something to eat. Food is excellent and you'll be tempted to sit here a while, just enjoying the setting.

When you're finally ready to leave, head back to Mervell Dean Road and turn left. Follow to the end and turn right onto MD 235 north, which will become MD 5 north as you head back to Washington, DC, and the Capital Beltway.

IN THE AREA

Accommodations

Executive Inn & Suites Park Avenue, 41655 Park Avenue, Leonardtown, 20650. Call 301-475-3000. Web site: www.execinnparkave.com.

Island Inn & Suites, 16810 Piney Point Road (MD 249), St. George Island, 20674. Call 301-994-1234. Web site: www.stgeorgeislandinnand suites.com.

Attractions and Recreation

Buzzy's Country Store, 12665 Point Lookout Road, Scotland, 20687. Call 301-872-5430. A great place to see where the locals hang out. You can also buy a drink or snacks. Web site: www.buzzyscountrystore.com.

Historic St. Mary's City, 18751 Hogaboom Lane, St. Mary's City, 20686. Call 240-895-4990 or 800-762-1634. First capital of Maryland; predated Annapolis. The site is now an outdoor history museum and archaeological park. Located on MD 5 and Rosecroft Road south of Leonardtown. Web site: www.stmaryscity.org.

Hole in the Wall Tavern, 24702 Sotterley Road (at Old Three Notch Road), Hollywood, 20636. Call 301-373-3838. A dive, but a dive with personality. Beer in bottles only—no drafts. Pool tables. And a horseshoe setup out back with stands for viewing; it's recommended that you bring your own horseshoes. Don't come hungry, for all they have besides alcohol are chips and Slim Jims.

Old Jail Museum, Courthouse Drive, Leonardtown, 20650. Call 301-475-2467. Built in 1858, this jail held many runaway slaves after they were captured. Open Wed. through Fri. Web site: www.stmaryshistory.org.

Patuxent River Naval Air Museum, 22156 Three Notch Road (MD 235), Lexington Park, 20653. Call 301-863-7418. Naval aviation museum is inside, with several historic planes and helicopters parked outside. Closed Mon. Web site: www.history.navy.mil/museums/paxmuseum /index.htm.

Piney Point Lighthouse, Museum and Historic Park, 44720 Lighthouse Road, Piney Point, 20674. Call 301-994-1471. Web site: www.co .saint-marys.md.us/recreate/museums.

Point Lookout State Park, 11175 Point Lookout Road, Scotland, 20687. Call 301-872-5688. Strategic location at the mouth of the Potomac River and the Chesapeake Bay. There's a pretty view, and one that's quite dramatic, with a lighthouse on the edge of the peninsula, plus camping and a fishing pier. Web site: www.dnr.state.md.us/publiclands/southern/point lookout.html.

Sotterley Plantation, 44300 Sotterley Lane (MD 245), Hollywood, 20636. Call 301-373-2280. This former tobacco plantation, with its steamboat landing on the adjacent Patuxent River, used slaves to work the fields and other jobs. Tours of the mansion provide insight into the times when this plantation thrived. There's an 1830s slave cabin on the property and elaborate gardens. Guided tours May through Oct; closed Mon. Self-guided tours year-round. Web site: www.sotterley.org.

Dining/Drinks

Island Bar & Grill, 16810 Piney Point Road (MD 249), St. George Island, 20674. Call 301-994-9944. Web site: www.stgeorgeislandinnand suites.com.

Clarke's Landing Restaurant, 24580 Clarke's Landing Lane, Hollywood, 20636. Call 301-373-8468. Great Maryland crab soup and crabcakes. Steamed crabs on the deck or porch in season. Web site: www.cl restaurant.com.

Kevin Thompson's Corner Kafe, 41565 Lawrence Avenue, Leonardtown, 20650. Call 301-997-1260. Excellent local dining, with specialty in seafood. The vegetables are a surprising treat. Maryland crab soup is spicy and delicious. Open for lunch Mon. through Sat. and dinner Wed. through Sat.

Other Contacts

St. Mary's County Division of Tourism, 23115 Leonard Hall Drive, P.O. Box 653, Leonardtown, 20650. Call 301-475-4200, ext.1404, or 800-327-9023. Web site: www.visitstmarysmd.com.

Boats tied up in front of Tiki Bar on Solomons Island

CHAPTER 6

Solomons Island and Points North

Exploring Boating and Beaches on the Bay's Western Shore

Estimated length: 130 miles
Estimated time: A weekend

Getting there: From I-495 (Capital Beltway) outside Washington, DC, take MD 5 south to MD 235 south. Then take MD 4 over the bridge across the Patuxent River to Solomons Island. If you're coming from Annapolis, head south on MD 2/4; it becomes South Solomons Island Road and leads you right onto the island.

Highlights: Fishing village of Solomons Island where the Patuxent River enters the Chesapeake Bay. Calvert Marine Museum and Drum Point Lighthouse. Marinas, restaurants, bars, and outdoor recreation. Fossil-rich beaches at Calvert Cliffs and Bayfront Park. Hiking and collecting shark's teeth. Victorian-style resort towns of Chesapeake Beach and North Beach. Railroad memorabilia and boutique shops.

Maryland is surrounded and defined by several bodies of water—the Atlantic Ocean, the **Chesapeake Bay,** and numerous large rivers, including the Potomac (which helps define Washington, DC), Patuxent, Severn, Susquehanna, Choptank, Nanticoke, and Patapsco Rivers. Most are named for the Indian tribes that once inhabited the lands.

Yet with all this water and the corresponding miles upon miles of shoreline, when people think of going to a Maryland beach, their automatic reaction is usually to think of the Eastern Shore, specifically Ocean City on the Atlantic coast. While that is an appealing option, there are also many charming places on the less-visited **Western Shore** of the Chesapeake Bay, some of which we are going to explore on this trip.

This area is much less congested than the Eastern Shore, especially in the summertime. On Maryland's Western Shore, you'll find the peaceful enjoyment that comes with being close to the water, and close to those who earn their living on the water, without the hordes of people flocking to the Atlantic Ocean on hot summer days.

Solomons Island is regularly frequented by many who live or work in Washington, DC, and its close-in Maryland suburbs, but it is an unknown treasure to many others throughout the region. Many of the people who live in Southern Maryland, which is where we're headed, even commute to DC for work if they don't work on a farm or on the water. These are long commutes, but the population is sparse, so there's not too much traffic until you get closer to DC.

Of all the historic waterfront towns and villages on the Western Shore, Solomons, also known as Solomons Island, is perhaps the most charming, and as such it has been built up quite a bit in recent years, becoming a popular escape for those who know about it. Solomons is actually a deep-water port located at the end of two peninsulas formed by the **Patuxent River, Mill Creek,** and **Back Creek;** all three bodies of water run into the Chesapeake Bay within sight of Solomons.

A bridge linking the island to the mainland on the west ascends and descends abruptly; this is **Governor Thomas Johnson Memorial Bridge,** which crosses the Patuxent River. Due to its abrupt height and descent, some call it the "Crazy Bridge." There are efforts afoot to replace it, but such efforts generally move slowly.

Once there, you can drive around Solomons, you can walk, and you can bike. There's so much to enjoy on the island and in the surrounding area that you'll want to spend the night. Allow plenty of time between activities so you can enjoy the leisurely atmosphere. Locals take their boats and their seafood seriously, but it is a relaxed seriousness.

Solomons Island was an active oyster shucking and packing center for more than a century, starting in the late 1800s. This fit the island well, for watermen here had been earning a living harvesting the Chesapeake Bay for more than three hundred years, subject of course to the whims of the weather. Now those professions are in jeopardy as environmental issues threaten the blue crabs and oysters that were once so prevalent. This makes tourism more important; recreational fishing and boating have also filled in some of the gaps.

Efforts are also under way to restore the health of the Bay, with some success where blue crabs are concerned. However, the current extreme reduction in the number of oysters poses a serious threat, for oysters actually purify water, thus counteracting pollution.

Fewer crabs and oysters, while affecting the economy and the environmental ecosystem, do not in any way diminish the beauty of the water and the shoreline. And luckily, there's no absence of seafood in the local restaurants.

After you cross the bridge, or come south on MD 2/4, and arrive in Solomons, you'll want to check in to a place to stay. There are three hotels on the island, the largest of which is the **Holiday Inn.** If bed & breakfasts are more your style and you can find an available room in one of the three homes with rooms to rent, that is also an option for staying in this charming boating community. Check in to your accommodations and head out for a walk or bike ride. There are also charter boats for fishing and sightseeing, as well as kayaks and canoes for rent.

You should plan to spend at least one night right on Solomons Island, for the lively entertainment, restaurants, and bar scene are inviting in a laid-back, casual waterside way. Solomons is a magnet for those who want a lively getaway, who want to charter a fishing boat, eat scrumptious seafood, and have an all-around great time.

This fishing village got its name from Isaac Solomon, the Baltimore businessman who started the first oyster-shucking cannery on the island right after the Civil War. Then prior to the Normandy invasion during World War II, U.S. servicemen practiced amphibious landings at Solomons. Now marinas abound, along with charter boats, fisherman, and many from the Washington, DC, area looking for a respite from urban life.

Recreation is a large part of the area around this part of the bay; yet there is also a great deal of history. The British attempted to use the Chesapeake Bay in several wars against the American "colonists." Of particular relevance to Solomons is the British invasion in the summer of 1814 toward the end of the War of 1812.

In fact, British Navy ships actually blockaded the Chesapeake Bay. The

Calvert Marine Museum in Solomons, in addition to documenting and describing the plentiful nature of the marine ecosystems and how they sustained generations of watermen and their families, also features many war artifacts. At the museum you can also see a live diamondback terrapin turtle, which is the official state reptile. (Testudo, the official school mascot for the University of Maryland in College Park, is also a terrapin.)

The Calvert Marine Museum is superb, a perfect place for both adults and children. It's a dramatic and entertaining museum, with boats and crabs and other subjects covered in equally entertaining fashion. Young kids are particularly fascinated by the fish tank and the turtles.

Outside, the **Drum Point Lighthouse** is a picturesque sight. It was decommissioned in 1962 and moved 3 miles to its current location in Solomons. Drum Point Lighthouse is a screwpile cottage-type lighthouse, one of only three such lighthouses remaining from the original 45 of that type that once served the Chesapeake Bay at the beginning of the 20th century.

As you walk or bike around downtown Solomons, you'll find that people are friendly, the scenery is delightful, and everyone is in a relaxed Caribbean kind of mood. The locals are watermen through and through, with generations in their families making their living near or on the water. Solomons was a quiet, sleepy place but that has changed. In the height of the summer season, you'll need reservations in advance, especially on weekends.

If you like art but want to spend time outside, get in your car and go north on MD 2/4. Turn right on Dowell Road for about a half mile to the entrance of **Annmarie Garden Sculpture Park & Arts Center.** Trails in the woods, with outdoor sculptures scattered around, combine to make for an interesting walk. Annmarie is connected with the Smithsonian and admission is free.

If you're a lighthouse enthusiast, besides the Drum Point Lighthouse, which you've already seen at the Calvert Marine Museum, you can follow Solomons Island Road a mile or so north and turn right onto Cove Point Road, which will turn into Lighthouse Boulevard and dead-end at the **Cove Point Lighthouse.** Built in 1828, the 40-foot brick tower contains an iron lantern with a lens manufactured in Paris in 1897. This lighthouse is the oldest continuously working lighthouse in the state. There's also a 1901 fog signal building and a light keeper's home in disrepair.

Retrace your route back down Lighthouse Boulevard and take Solomons Island Road back onto the island. For dinner, there are many good options, one of which is **Stoney's Solomons Pier.** Crabcakes are good at Stoney's, as are the steamed hardshells if they are in season. After dinner, you'll want to stay and have at least one drink at the adjacent bar. Not far away, there's also the **Tiki Bar,** a popular hangout for boaters over many decades and still going strong.

In the morning, you'll have breakfast at your B&B. Or if you're staying at one of the hotels, consider eating breakfast at the **Captain's Table,** a popular local favorite on the waterfront near the Holiday Inn.

After breakfast, get back into your car and start traveling up the coast in the direction of Baltimore, just as the British did in 1814 during the "Second War for Independence" otherwise known as the War of 1812. You'll soon come across H.G. Trueman Road (MD 765). Take the road 1.3 miles to the entrance of **Calvert Cliffs State Park** near the town of **Lusby.** Miles of cliffs along the edge of the park are more than 15 million years old, formed when all of the land in Southern Maryland was covered by a warm, shallow sea.

At Calvert Cliffs, you can hike to the beach (about 2 miles from the parking lot), where the cliffs are eroding, setting free fossils and sharks' teeth, which can be collected along the water, along with more recent shells. It's certainly a collector's delight and children love to find these free-for-the-taking treasures.

After you spend some time at Calvert Cliffs, you'll probably be ready for lunch. The **Frying Pan** is a restaurant right on H.G. Trueman Road in Lusby. It looks like a hole in the wall, but it is a local favorite and is really quite good. The food is inexpensive and tasty, and the servers take good care of you.

After a quick lunch (or a slow one if you feel like it), continue north on MD 2 until you reach the small town of **St. Leonard.** On July 4, 1814, the British burned the wharves and warehouses, destroying the town before heading to Washington, DC, which they proceeded to burn on August 24, 1814.

Now there's a homemade candy store in St. Leonard named **Sweet Dreams** and a mini–antiques mall, **Chesapeake Market Place,** with friendly vendors and assorted treasures. There's also a **Polling House and Garden of Remembrance.** Built in 1926, the polling house was used until 1974. The garden honors people considered important by area residents, as well as two First Ladies, Mrs. John Quincy Adams and Mrs. Zachary Taylor, who had family ties in St. Leonard.

If you're a real history buff, you can detour to **Benedict,** a riverside town on the Patuxent River where the British sought a landing spot from which to invade Washington, DC. To get there, follow MD 2/4 north and take MD 231 west; you'll cross the Patuxent River into Benedict. After this detour, take MD 231 east until you are back where it intersects with MD 2/4 and head north once more.

If you skip Benedict, take MD 2/4 directly from St. Leonard to the town of **Prince Frederick.** You can't miss this town, because of its many box stores and fast-food outlets. If you need anything from a drugstore (or other retailer), you might want to get it here.

From Prince Frederick, continue following Solomons Island Road (MD 2/4) north until you reach Plum Point Road (MD 263). Take a right onto Plum Point approximately 4.1 miles to a split in the road. Bear left onto Bayside Road (MD 261) and drive about 5.7 miles.

You'll pass through rural landscape until you enter the beach town of **Chesapeake Beach.** At the first traffic light, turn right into the **Chesapeake Beach Resort & Spa,** a Victorian-style hotel and marina where you'll smile just pulling up to the cheerful facade. This resort is a perfect place to spend another night; it's comfortable without being fancy.

The Victorian style of the resort is appropriate, for this became a popular beachfront town when the railway station was built in 1897 by developer Otto Mears to provide transportation to and from Washington, DC. The original railroad depot is now the tiny **Chesapeake Beach Railway Museum** right outside the front door of the hotel. You'll find the display interesting, for some Chesapeake Beach enthusiasts once had dreams of the town rivaling Atlantic City, New Jersey. That was not meant to be, of course, and this museum helps explain the reasons why not.

North Beach, a kind of sister town to Chesapeake Beach, is less than a half mile away, with a half-mile-long boardwalk, a bike path, and a public fishing pier plus excellent restaurants and boutique shops.

In addition to train passengers arriving from Washington, DC, many visitors arrived in Chesapeake Beach by steamboat from Baltimore. However, once the Great Depression hit in the 1930s and the local Otto Mears railroad failed in 1935, the popular seaside resort fell into disrepair and disuse, at least partially. Construction of the **Chesapeake Bay Bridge** was

You can go fossil hunting for sharks' teeth at "Brownie's Beach" (Bayfront Park).

another blow to Chesapeake Beach and North Beach, for travelers began flocking to the Eastern Shore and **Ocean City.** Although that trend continues, these Western Shore beach towns are being rediscovered. Still, they are much quieter and more remote than Ocean City, which is part of their charm.

If you skipped Calvert Cliffs or if you fell in love with fossil hunting, you can also do some of it on a smaller scale at the beach that locals call "Brownie's Beach," though it's really **Bayfront Park.** If you sift through the sand a while, you should at least find some tiny sharks' teeth. Views are limited at Bayfront Park, but if you like a sense of isolation, this is a good spot.

View from Chesapeake Beach Resort & Spa

Besides going to the beach, you can rent kayaks and bikes at **Paddle or Pedal,** take the kids to **Chesapeake Beach Water Park,** and play slots at **Rod N Reel** or **Traders Seafood Steak & Ale.** Among the lovely boutique shops and restaurants in North Beach is the **Westlawn Inn,** open for dinner with superb cream of crab soup, fried asparagus, and fresh rockfish—Maryland's favorite local fish. There's also a spa at the Chesapeake Beach Resort & Spa, so you can get a massage before heading home.

To complete the trip, continue driving north on MD 2 if you started out from Annapolis. (MD 2 splits from MD 4 shortly after the spot where you turned off onto MD 263 toward Plum Point, so it's just MD 2 going north from Chesapeake Beach.) Alternately, head back to the DC suburbs via MD 260 west to MD 4 north to get to the Capital Beltway (I-495).

IN THE AREA

Accommodations

Back Creek Inn Bed & Breakfast, 210 Alexander Lane, Solomons, 20688. Call 410-326-2022. Great views of Back Creek. Web site: www.back creekinnbnb.com.

Blue Heron Inn Bed & Breakfast, 14614 Solomons Island Road, Solomons, 20688. Call 410-326-2707. Beautiful contemporary-style B&B. Owner is a chef, so breakfast is a treat. Web site: www.blueheronbandb .com.

Chesapeake Beach Resort & Spa, 4165 Mears Avenue, Chesapeake Beach, 20732. Call 410-257-5596 or 866-312-5596. A definite nostalgic, turn-of-the-20th-century feel to this resort will make everyone in the family smile. There are 72 water-view rooms with flat-screen TVs and high-speed Internet. There's a marina right outside so you can arrange to go charter fishing and there's a good restaurant, the Rod N Reel, with a bar and many slot-type machines. Web site: www.cbresortspa.com.

Comfort Inn Beacon Marina, 255 Lore Road, Solomons, 20688. Call 410-326-6303 or 800-228-5150. Smaller than the Holiday Inn, the Comfort Inn has 60 rooms and is also situated in the heart of Solomons next to a marina. You can park your car and walk to many of the attractions. Web site: www.comfortinn.com.

Hilton Garden Inn Solomons, 13100 Dowell Road, Solomons, 20688. Call 410-326-0303. The newest of the three hotels on the island, the Hilton is away from the center of the island and requires a car to get around, though the island is small and easily navigated. Web site: www .hilton.com.

Holiday Inn Solomons Conference Center & Marina, 155 Holiday Drive, Solomons, 20688. Call 410-326-6311. The largest hotel on the island with 320 rooms, among which just 4 of the rooms have pull-down beds, which might seem like fun but are rather uncomfortable. However, the 316 other rooms have regular beds that are modern and comfortable. Many of the rooms even have lovely waterfront views of the marina, and

the hotel is within walking or biking distance of the historic downtown.
Web site: www.holiday-inn.com.

Solomons Victorian Inn Bed & Breakfast, 125 Charles Street,
Solomons, 20688. Call 410-326-4811. Great location but needs some renovation. Web site: www.solomonsvictorianinn.com.

Attractions and Recreation

Annmarie Garden Sculpture Park & Arts Center, 13480 Dowell
Road, Dowell, 20629. Call 410-326-4640. An outdoor sculpture garden
and 30-acre retreat affiliated with the Smithsonian. It is a peaceful place
to enjoy art in the outdoors. Open daily year-round. Free admission. Web
site: www.annmariegarden.org.

Bayfront Park, one-half mile south of Chesapeake Beach. Call 410-257-
2230. Otherwise known as "Brownie's Beach" by locals. You'll need to
park your car and follow the path about a half-mile hike to the beach.
Admission fee Memorial Day through Labor Day; free off-season.

Bayside History Museum, 9006 Dayton Avenue, North Beach, 20714.
Call 410-495-8386. Old-fashioned slot machines and other local memorabilia make this an interesting place to stop for 20 minutes or so. It's free
and right in town, so it's easy to do. The exhibits document the little
beach community that saw its heyday when a railway ran between the
beach and Washington, DC, and before the Chesapeake Bay Bridge was
built. Open Wed. through Sun. from Memorial Day through Labor Day.
Web site: www.baysidehistorymuseum.org.

Beach Trolley, operating between the towns of Chesapeake Beach and
North Beach, as well as to Deale and Dunkirk. A ride is just 25 cents. The
trolley makes a full loop every 2 hours with about 50 stops in between.
Runs Fri. through Sun. between Memorial Day and Labor Day. Riders
under 12 must be accompanied by an adult. Web site: www.beachtrolley
association.org.

Calvert Cliffs State Park, 9500 H.G. Trueman Road (MD 765), Lusby,
20657. Call 301-743-7613. You park your car and hike through the state

park. If you want to walk the beach to search for fossils, it's about a 2-mile jaunt from where you park. The cliffs rise more than 100 feet above the water. Open daily year-round from sunrise to sunset. Web site: www.dnr .state.md.us/publiclands/southern/calvertcliffs.html.

Calvert Marine Museum, 14200 Solomons Island Road, Solomons, 20688. Call 410-326-2042. A well-thought-out museum that is interesting for both adults and children. Live Maryland terrapin turtles and fish, plus a dramatic shark fossil and information on the indigenous blue crab and other marine animals. There's a lighthouse outside that has been moved to the site and is no longer operating, but it is delightfully picturesque. Open daily. Web site: www.calvertmarinemuseum.com.

Chesapeake Beach Railway Museum, 4155 Mears Avenue, Chesapeake Beach, 20732. Call 410-257-3892. Small but interesting, this tiny museum is on the grounds of the Chesapeake Beach Resort & Spa, right outside the front door of the hotel. Free admission. Web site: www.cbrm.org.

Chesapeake Beach Water Park, 4079 Gordon Stinnett Avenue, Chesa-peake Beach, 20732. Call 410-257-1404. Web site: www.chesapeakebeach waterpark.com.

Chesapeake Market Place, 5015 St. Leonard Road, St. Leonard, 20685. Call 410-586-1161. Antiques, and upstairs there's a coin shop only open on Sunday. At one time this was a lumberyard; now you can discover someone else's treasures and collectibles. Open Wed. through Sun. Web site: www.chesapeakemarketplace.com.

Cove Point Lighthouse, 3500 Lighthouse Boulevard, Lusby, 20657. This is the oldest continuously working lighthouse in the state, with views on the Bay of an offshore oil pipeline. Rather remote and basic, but if you like lighthouses, you should stop here. National Register of Historic Places.

Drum Point Lighthouse, 14200 Solomons Island Road, Solomons, 20688. Constructed in 1883 at Drum Point to mark the entrance to the Patuxent River, this lighthouse was decommissioned in 1962 and recon-structed outside the Calvert Marine Museum, about 3 miles from its orig-inal location. National Register of Historic Places.

Island Trader Antiques, 225 Lore Road, Solomons, 20688. Call 410-326-3582. Closed Wed.

North Beach Public Beach, Fifth Street & Bay Avenue, North Beach. There's a small boardwalk, just a little over half a mile long. The free beach is open 6 AM–10 PM. It's a great, not-too-crowded place to get sand between your toes and water in your hair.

Paddle or Pedal, 4055 Gordon Stinnett Avenue, Chesapeake Beach, 20732. Call 410-991-4268. Web site: www.paddleorpedal.com.

Patuxent Adventure Center, 13860 Solomons Island Road, Solomons, 20688. Call 410-394-2770. One of the many places to rent bicycles and kayaks on the island. Open year-round every day but Mon. Web site: www .paxadventure.com.

St. Leonard Polling House, St. Leonard Road, St. Leonard, 20685. Located in a quiet little park. You can drive by this historic polling house or stop for a quiet break.

Sweet Dreams Candy Shoppe, 4902 St. Leonard Road, St. Leonard, 20685. Call 410-610-3669.

Dining/Drinks

Boomerangs Original Ribs, 13820 H.G. Trueman Road, Solomons, 20688. Call 410-326-6050. Great place for dinner if you love barbecued ribs. Other goodies include beer battered onion rings. Web site: www.love ribs.com.

CJ's Back Room, 12020 Rousby Hall Road, Lusby, 20657. Call 410-326-1125. Lusby's answer to Cheers, this local hangout offers great pub food. Open daily.

The Frying Pan, 9895 H.G. Trueman Road, Lusby, 20657. Call 410-326-1125. A seeming hole in the wall, this restaurant is friendly, the food inexpensive and tasty, and the atmosphere comfortable. Locals love it. Open daily.

Rod N Reel, 4160 Mears Avenue, Chesapeake Beach, 20732. Call 410-257-2735. Although this is a large place, obviously a hotel restaurant and

Drum Point Lighthouse outside the Calvert Marine Museum

slightly lacking in ambience, it is convenient and serves good food, including decent crabcakes. Also has slot-type machines and a bar. Open daily. Web site: www.cbresortspa.com.

Ruddy Duck Brewery & Grill, 13200 Dowell Road, Dowell, 20629. Call 410-394-3825. Microbrewery and restaurant. Open daily. Web site: www .ruddyduckbrewery.com.

Solomons Island Winery, 515 Garner Lane, Lusby, 20657. Call 410-394-1933. Solomons Island is, after all, a drinking kind of place. Web site: www.solomonsislandwinery.com.

Stoney's Solomons Pier, 14575 Solomons Island Road, Solomons, 20688. Call 410-326-2424. This lively, popular place is large but it is made up of smaller rooms, so it has a friendly, intimate feel. Service is pleasant, crabcakes are delicious, and if you order steamed hardshells, other patrons will ask you about the size and weight and taste, so be prepared to be friendly. Eating crabs and drinking beer are, after all, a communal activity. The bar is crowded on weekends, but there's always room to squeeze in. Open daily. Web site: www.stoneysseafoodhouse.com.

Tiki Bar, 81 Charles Street, Solomons, 20688. Call 410-326-4075. A bar where boaters hang out. Popular for decades, it continues to be crowded, especially on summer weekends. Open mid-Apr. through Oct. Web site: www.tikibarsolomons.com.

Traders Seafood Steak & Ale, 8132 Bayside Road, Chesapeake Beach, 20732. Call 410-257-6126. Another option for food in this lovely beach-side town. The big draw is the slot-like "electronic bingo" machines, which are allowed here by the state as long as they are not called slots. This is a more upscale atmosphere in which to gamble compared with the rooms of "faux slots" at the Rod N Reel in the Chesapeake Beach Resort & Spa. Serves breakfast, lunch, and dinner. Dining and gaming daily. Web site: www.traders-eagle.com.

The Westlawn Inn, 9200 Chesapeake Avenue, North Beach, 20714. Call 410-257-0001. Upscale restaurant with excellent food and wine list. Saturday night there's live music and there's a fun bar where you can dine if you prefer a more casual atmosphere than in the dining room. Dinner Tues. through Sun. and brunch Sat. and Sun. Web site: www.westlawn inn.com.

Other Contacts

Calvert County Tourism, Department of Economic Development, 175 Main Street, Prince Frederick, 20678. Call 410-535-4538 or 800-331-9771. Web site: www.ecalvert.com.

Chesapeake Beach, P.O. Box 400, Chesapeake Beach, 20732. Web site: www.chesapeake-beach.md.us.

Solomons Business Association, Web site: www.solomonsmaryland .com.

Town of North Beach, 8916 Chesapeake Avenue, North Beach, 20714. Call 301-855-6681. Web site: www.ci.north-beach.md.us.

The Awakening, *a sculpture of a giant in the sand at National Harbor*

CHAPTER

7

Southern Tobacco Country

Tracing the Escape Route of John Wilkes Booth

Estimated length: 90 miles
Estimated time: Overnight (not including detour to Booth's birthplace and gravesite)

Getting there: From I-495 (Capital Beltway) near Silver Spring, travel south (inner loop) toward Andrews Air Force Base and Alexandria, Virginia. At the split for local and express lanes, stay to the right (local lanes). Take exit 2A toward National Harbor (the last exit in Maryland before crossing the Potomac River into Virginia). Follow the signage to National Harbor. Once there, turn right on St. George Street and follow signs to Gaylord National's self-park garage.

Highlights: Places John Wilkes Booth went for assistance and to hide after assassinating the president of the United States: a tavern, a cemetery, a doctor's house, a swamp, the Potomac River.

We'll start this trip from **National Harbor,** which is a new community of hotels, retail stores, condominiums, and a marina on the Maryland side of the Potomac River just southeast of Washington, DC. If you're looking to trace the hunt for a presidential assassin by day and enjoy relative luxury by night, this is the place to stay. The **Gaylord National Resort & Conven-**

tion Center is the largest of several hotels at National Harbor. Check in and relax before you start contemplating the escape route of John Wilkes Booth, infamous assassin.

Another logical place to start would be at **Ford's Theatre** in the nation's capital, where the fatal deed was actually done. But for our purposes, we will examine the places in Southern Maryland where Booth sought aid during his attempted escape from federal authorities who were determined to catch him.

For context, the Civil War was virtually over with the surrender by Robert E. Lee and his army, but that in no way ended the controversy over the issues that had caused the war in the first place. President Abraham Lincoln had received death threats before. It was a volatile time.

Many people who grow up in Maryland think of themselves as Northerners. After all, the state remained with the Union during this nation's devastating Civil War. Virginia to the South went with the Confederacy. And many believe that the Mason-Dixon Line, reputed to separate North and South, however arbitrary its logistics might be, must separate Maryland and Virginia. But this belief, held by many, is patently false.

In truth, the Mason-Dixon Line separates Maryland from Pennsylvania in the north and Delaware to the east. And within the state itself, the residents of Maryland were divided on issues of states' rights and slavery in the mid-1800s. In fact, the state may have seceded from the Union along with its Southern neighbors had the federal military not stepped in and controlled the city of Baltimore and the state's forts and waterways, and arrested some pro-Confederate members of the legislature prior to a vote deciding on whether to stay or go from the Union.

Perhaps it is no wonder that a native Marylander, John Wilkes Booth, would be the one to bring down the symbol of all that opposed the Confederacy—President Abraham Lincoln. And given the Southern sympathizers in the state, especially in the southern counties, it's no wonder that Booth received aid and comfort in his attempted escape...until, that is, he crossed the Potomac into Virginia. Tired of war and facing the harsh realities of their recent defeat, Virginians gave less effective support to Booth, ultimately leading to his death at the hands of armed federal agents.

Abraham Lincoln, the Civil War, and John Wilkes Booth capture the imagination of many. While most view Booth as a villain, he is no less interesting, no less a fascinating character. And his story is uniquely one of Maryland.

While you're at Gaylord National Hotel, relax, do some shopping or swimming, and have dinner at the **Old Hickory Steakhouse.** If you like cheese, you'll love the artisanal cheeses they keep in their "cheese cave." National Harbor was developed after much controversy, just outside DC city lines in Maryland. Some of the rooms at Gaylord overlook the Potomac River, Washington, and Alexandria, Virginia, so it's an interesting planned development, though a bit contrived.

While at National Harbor, make sure to take a look at sculptor J. Seward Johnson's famous *The Awakening.* This huge sculpture was previously located at East Potomac Park in DC, but was moved to National Harbor. It's in a few pieces—a giant's hands and face placed in a large sandbox of sorts, or a teeny tiny beach, providing somewhere for young children to play while their parents watch over them and snap photos of the kids crawling around the sculpture.

There's a small marina, as well as views of Washington and Alexandria, Virginia, across the Potomac River. It is interesting to note that the land where Washington is located was once part of the state of Maryland; but it was ceded to the federal government so that the nation's capital could be established. The Potomac River actually belongs 100 percent to Maryland, unlike many other "shared" rivers that divide or form the boundary between two states. Even so, various court decisions have granted certain water rights to Virginia; but technically the river belongs entirely to Maryland.

The route of the assassin began, of course, at Ford's Theatre in the nation's capital. Once the deed was done on April 14, 1865, Booth jumped on his horse in back of the theater and galloped away through the streets of DC and across a bridge into Maryland. He headed for **Surrattsville,** a small town near modern-day **Clinton.**

Keep in mind that 12 days elapsed between Booth's shooting of the president and his being shot by a federal agent at a barn in Virginia. For several days he hid in a pine thicket near a swamp, and in some areas, his exact route is unknown. However, in one day, you can see the main places Booth went in Maryland and get an idea of his escape route with conspirator David Herold. This trip is not meant to be an exact re-creation but rather a representative one through what was, at the time, a tobacco-producing area.

During the Civil War, tobacco plantations in Southern Maryland ranged from large ones with 100 and more slaves to more middle-class plantations with fewer than 10 slaves. There were also some farmers who couldn't afford to buy a slave and who worked the land themselves. After the war, many of the slaves returned to work as tenant farmers for the former slaveholders.

To start tracing Booth's escape route, take I-95 to exit 7 and go south on Branch Avenue (MD 5) to Woodyard Road (MD 223) west. Turn right and go 1 mile to the second light, where you'll turn left onto Brandywine Road. The **Surratt House** is on your left.

The Surratt House is certainly one of the crucial and most interesting aspects of this tour. It played a huge part in the assassination plot and it is also a fascinating bit of Americana. Built in 1852 as a home for the Surratt family, it was also a tavern. Like many taverns in those days, this one was located at a crossroads and had multiple uses. Besides being a home and tavern, it was a public dining room and a hotel of sorts for traveling gentlemen. A livery stable and blacksmith shop assisted travelers on horseback.

In 1854, a post office was added to the tavern. Mr. Surratt was the postmaster and the town became Surrattsville, which was common parlance for a town with a post office—the town name was related to the postmaster, who served a vital community function. The Surratt House was also an official polling place, just one of the political roles it played.

The guided tour at the Surratt House is well done and worth taking. There's also a small gift shop, and the Surratt Society, which supports the Surratt House, offers all-day bus tours of the escape route; however, there is a limited number of these bus tours during the year and they sell out almost immediately. You can certainly do it yourself and at your own pace, which is what this trip is about.

The Surratt House is where Booth stopped to retrieve weapons and supplies after assassinating Lincoln. Booth knew Mary Surratt through one of her sons, who was a Confederate courier. In fact, Booth recruited her son into his original plot to kidnap President Lincoln; another Surratt boardinghouse in Washington was the site of several conspiracy meetings.

The kidnap plan never materialized; instead Booth shot and mortally wounded the president, who died the next day. Afterward, Booth escaped Washington on horseback and stopped at the Surratt House in what is now Clinton, Maryland, for guns and other supplies needed for his escape.

Next on the escape route is the **Dr. Samuel A. Mudd House,** which is really a farmhouse. To get there, head south on Brandywine Road for 5 miles. Make a slight right to merge onto Branch Avenue (MD 5) south. Turn left at Mattawoman Beantown Road (MD 5) south. Turn left at Poplar

The Dr. Samuel A. Mudd House, where Booth was treated for his broken leg

Hill Road and follow to a slight right turn at Dr. Samuel Mudd Road.

The Mudd House is interesting, but less so than the Surratt House. The tour covers the Mudd family in great detail, but not much is said about John Wilkes Booth's infamous visit here to have his leg tended to after injuring it at Ford's Theatre.

Next you'll want to visit the **Old Bryantown Tavern** where Confederate agents met with Booth before the assassination. To get there, turn right on Dr. Samuel Mudd Road and go about 1.5 miles. Turn right on Bryantown Road (MD 232). Drive 3 miles, cross MD 5, and turn right onto Trotter Road. The second house on the right is the now privately owned Old Bryantown Tavern.

One mile south of the Bryantown Tavern, on the east side of MD 232, is **St. Mary's Catholic Church,** where Dr. Mudd and Booth first met in 1864. Dr. Mudd is buried in the cemetery; when you face the church, Mudd's grave is on the left in the front row.

MORE TO THE STORY

If you want to take a detour to see where John Wilkes Booth lived as a child, his family home is in the town of Bel Air in another part of the state. To get there, take I-95 north to exit 80 (MD 543/Churchville). Turn left off the exit onto MD 543 and travel 5.5 miles to Churchville Road (MD 22). Turn right onto Churchville Road and travel for 0.6 mile to Tudor Lane on your left. Follow Tudor Lane to the end, pass through the stone gates, and continue to the end of the driveway. (Drive slowly because it is a residential area with children.)

Tudor Hall is interesting to visit with its pretty, well-tended grounds. Booth's father and brother were also famous actors, so drama ran in the family. But no one expected such a drama as the assassination of the president. Tours of the house are given sporadically and require reservations.

Tudor Hall, where John Wilkes Booth lived as a child

If you really want to see the story to its conclusion, you can visit Baltimore's **Green Mount Cemetery,** where the former actor and presidential assassin is buried. Besides the infamous Booth, many of Baltimore's notable philanthropists, such as Enoch Pratt, Johns Hopkins, and William and Henry Walters, are also buried there.

When buried, John Wilkes Booth was put into an unmarked grave to prevent vandalism. Its exact location is unknown; however it is in the vicinity of the Booth family plot. Green Mount Cemetery is in a rough, dicey section of downtown Baltimore, so it is not a recommended stop except for the "diehard." Booth was actually buried in the plot on June 22, 1869, about four years after he was killed. Prior to that, his body was buried in a storage room at the Old Penitentiary and later moved to a warehouse at the Washington Arsenal. Edgar Allan Poe was also buried three times in Baltimore, but for other reasons.

Some conspiracy theorists believe that Booth actually escaped and a look-alike was killed in his place. However, most historians believe Booth died at the Virginia barn.

From the church, turn left on Oliver's Shop Road and drive for almost 5 miles. Turn right on Charles Street (MD 6) west. Travel about 4 miles past a sign for **Zekiah Swamp,** which Booth and Herold crossed, and turn left onto Bel Alton Newtown Road. On this road, you will pass by Rich Hill, the home of Samuel Cox, who directed the fugitives to a safe haven in a nearby pine thicket. As you cross a railroad track, the thicket can be seen on the horizon to your left. Turn left again at Crain Highway (US 301) for 1.5 miles and turn right at Popes Creek Road for about 1 mile.

Popes Creek is the local name for a little area along Popes Creek Road where there are a few crab houses. It is a quiet, unassuming place off-season, but in the summertime large crowds of people arrive by boat and car. Lots of these people monopolize tables at **Captain Billy's Crab House** eating steamed crabs. Many people illegally crossed the Potomac River into Virginia here during the Civil War, and this is where John Wilkes Booth also crossed on his fateful last trip. While you're here, have lunch or dinner.

> **THE OVERALL ESCAPE ROUTE**
> Ford's Theatre, Navy Yard Bridge, Surratt House Tavern in the village of Surrattsville (now Clinton), village of T.B. (named for former landowner Thomas Brooke), Dr. Samuel A. Mudd's home 3 miles north of Bryantown, Zekiah Swamp, home of Samuel Cox near present-day town of Bel Alton, pine thicket, Potomac River, Nanjemoy Creek, Potomac River, Rappahannock River in Virginia, tobacco barn on Richard Garrett's farm in Virginia.

To return, head back on Popes Creek Road and turn left onto US 301 north. Go left again on Mattawoman Beantown Road (MD 5) north and then back to I-495 (Capital Beltway).

IN THE AREA

Accommodations

Colony South Hotel, 7401 Surratts Road, Clinton, 20735. Call 301-856-4500 or 800-537-1147. Right down the road from the Surratt House. Web site: www.colonysouth.com.

Gaylord National Resort & Convention Center, 201 Waterfront Street, National Harbor, 20745. Call 301-965-2000. A lovely property seemingly in the middle of nowhere but actually rather close to Washington, DC, and just across the Potomac River from Alexandria, Virginia. Some rooms overlook the Potomac River. Superb steakhouse on-site as well as shopping in the little, newly created community of National Harbor. Pool, spa, shops, water taxis. Web site: www.gaylordnational.com.

Attractions and Recreation

The Awakening, Waterfront Beach, National Harbor, 20745. Formerly located in East Potomac Park in the nation's capital, this unusual sculpture by J. Seward Johnson, Jr. was moved and placed here in the sand at the little waterfront beach. It's a great place for small children to play (with supervision) and for photo ops.

Dr. Samuel A. Mudd House, 3725 Dr. Samuel Mudd Road, Waldorf, 20601. Call 301-274-9358 or 301-645-6870. After Lincoln's assassination, John Wilkes Booth showed up here to have Dr. Mudd set the broken leg he got when he jumped onto the stage from the president's box at Ford's Theatre. Open Wed., Sat., Sun., late Mar. through Nov. Admission fee. Web site: www.somd.lib.md.us/MUSEUMS/Mudd.htm.

Green Mount Cemetery, 1501 Greenmount Avenue, Baltimore, 21202. Call 410-539-0641. Large graveyard where John Wilkes Booth is buried. Main gate is at Greenmount Avenue and East Oliver Street. Web site: www.greenmountcemetery.com.

St. Mary's Catholic Church Cemetery, 13715 Notre Dame Place, Bryantown, 20617. Call 301-870-2220. Cemetery where Dr. Samuel Mudd, who treated Booth's leg after the assassination, is buried. Web site: www.stmarysbryantown.com.

Surratt House and Tavern (Museum), 9118 Brandywine Road, Clinton, 20735. Call 301-868-1121. John Wilkes Booth stopped here after shooting the president. For her part in the assassination plot, tavern owner Mary Surratt was the first woman executed by the U.S. government. Open Thurs. through Sun. except holidays; closed mid-Dec. through mid-Jan. Admission fee. Web site: www.surratt.org and www.pgparks.com.

Tudor Hall, 17 Tudor Lane, Bel Air, 21015. Call 443-619-0008. The Booth family home, off the beaten path, with lovely grounds, mature trees, and a pond, but the house tours are spotty, depending upon which guide you get. The house tour focuses too much on the house and family and not enough on Booth and his story. Still, a visit is pleasant and it rounds out the story. Web site: www.spiritsoftudorhall.blogspot.com.

Dining/Drinks

Captain Billy's Crab House, 11495 Popes Creek Road, Newburg, 20664. Call 301-932-4323. Great place to contemplate John Wilkes Booth's escape near here while enjoying the waterfront ambience and eating steamed hardshell crabs in season. Open for lunch and dinner; closed Mon. and in the wintertime. Web site: www.captbillys.com.

National Pastime Sports Bar & Grill, 201 Waterfront Street, National Harbor, 20745. Call 301-965-5500. Food and fun surrounded by sports memorabilia like the home plate from RFK Stadium and a statue of Babe Ruth, who was born in Maryland. Sporting events on flat-screen TVs, golf simulator, and casual food, including wings, chili, chicken noodle soup, hot dogs, burgers, and ribs. Open daily. Web site: www.gaylordnational .com.

Old Hickory Steakhouse, 201 Waterfront Street, National Harbor, 20745. Call 301-965-4000. Excellent steaks, of course, plus artisanal cheeses, wine cellar, and great views. Web site: www.gaylordnational.com.

Other Contacts

Baltimore Area Convention & Visitors Association, 100 Light Street, 12th Floor, Baltimore, 21202. Call 410-659-7300. Web site: www.baltimore .org.

Charles County Office of Tourism, 12480 Crain Highway, Newburg, 20664. Call 301-259-2500. Web site: www.meetcharlescounty.com.

Harford County Office of Tourism, 220 South Main Street, Bel Air, 21014. Call 410-638-3327. Web site: www.harfordmd.com.

Prince George's County Conference & Visitors Bureau, 9200 Basil Court, Suite 101, Largo, 20744. Call 301-925-8300 or 888-925-8300. Web site: www.visitprincegeorges.com.

Commercial buildings at Annapolis City Dock

CHAPTER

8

Annapolis— Sailors' State Capital

Tracking Political Intrigue in
Little-Known Places

Estimated length: 120 miles
Estimated time: Overnight

Getting there: From Baltimore, take I-695 east to I-97 south (exit on the left). I-97 merges with US 50 east toward the Bay Bridge. Stay on US 50 east about 1 mile and take a right at exit 24 to Rowe Boulevard (MD 70) and bear right toward Annapolis. Stay on Rowe Boulevard and go straight through two lights; the road will veer to the right, taking you to a third light. At this light, bear left onto Northwest Street. At Church Circle, travel halfway around the circle and turn right onto Duke of Gloucester Street. Turn left at Green Street and left again onto Main Street. Look for on-street parking or take a left onto Gorman Street and park in City Garage.

Highlights: State capital. Boats and hangouts for sailors. Eastport. U.S. Naval Academy. Points south along the Chesapeake Bay and local rivers. Great restaurants. Alex Haley monument at City Dock.

State capitals are often sleepy little cities when the state legislature is not in session, but not so **Annapolis.** College towns are often quiet when school is not in session, but not so Annapolis. With the **U.S. Naval Academy** and **Saint John's College,** Annapolis does qualify as a college town, but it is not typical. When the schools are not in session, Annapolis is still active.

That's because it has so much going for it. After all, it's a haven for sailors and those who like to hang out with sailors. So besides the inevitable political intrigue of a capital, there is the fascination of, and preoccupation with, the water and the weather and the vessels used to conquer both.

The downtown area reflects a sense of humor mixed with historical relevance. Two circles—State and Church—lend credence to the separation of religion and government established in early America. That is why so many came here. They stayed because this place is beautiful.

Annapolis was once the capital of the United States. Now it is the capital of the state and it is also a playground for those who love being near the water to enjoy its beauty and the wonderful food that comes from it.

People are friendly, in a small-town kind of way. You can stop the mailman and ask for the location of a florist; he will direct you through an alley off Main Street. Or you can wander into **Chick & Ruth's Delly,** a tradition in these parts. If you're an early riser, get there for breakfast so you can witness everyone in the place standing, putting their hands to their hearts, and reciting the Pledge of Allegiance to the U.S. flag.

Chick & Ruth's classic sandwiches are named after the clerk of the circuit court—whose office is just a few blocks up the street in the courthouse at Church Circle—and various congressmen, judges, and state senators, plus the governor, former governor, lieutenant governor, attorney general, police chief, county sheriff, and other politicos. To have a sandwich named after you at Chick & Ruth's is a rite of passage in Annapolis. Ted runs the place that his parents started. The main attractions are the food...and the rituals.

Annapolis is a charming seaport with colonial-era buildings mixed with more modern structures. At the **City Dock,** you have a great view of **Ego Alley,** which is where affluent boaters bring their yachts to show off. On the other end of the spectrum, and also at the City Dock, is a monument to Alex Haley, author of *Roots,* whose ancestor Kunta Kinte debarked from a slave ship on that very dock in 1767.

There's much to do in historic downtown Annapolis. There are many stores, restaurants, and bars and a good time is generally had by all. But if you want to get out in the country, take a drive to **Jimmy Cantler's Riverside Inn,** which was opened in 1974. This is an isolated restaurant destination, but it can be lively during the summer season. Food is adequate but not sensational—for that you'll have to go elsewhere. But this is certainly a picturesque location with a well-defined, quirky personality. It's a classic, true down-home Maryland experience on the water.

People with boats pull up, crabs are molting, and there is a lively crowd on Saturday nights. Cantler's is not fancy, but it is authentic and down-to-the-bones real. Cantler's is on Mill Creek, which looks like a river, for there's a lot of water. But technically it's a creek off the Severn River across from the Naval Academy. To find Cantler's, take MD 2 north from Annapolis. Turn right onto Ritchie Road straight through the four-way to St. Margaret's Road. Go right onto Browns Woods Road to a right on Forest Beach Road. Follow it to the end.

Of course, back in town, you're close to the Naval Academy, and you should make sure to find time to visit the campus. The crypt of Revolutionary War naval hero John Paul Jones is located beneath the main chapel. No visit is complete without a stroll through the U.S. Naval Academy campus, which is known as the Yard. When the midshipmen are around during the academic school year, the Yard is more lively than in the summertime. Still, this is a beautiful, historic campus filled with traditions and worthy of a visit anytime.

In Annapolis, people talk about who hangs out in which bars and with whom. Politicians, especially when the legislature is in session, are mixed in with the locals and the tourists, so it makes for a lively time. There are rumors of ghosts at some establishments and tall tales of beds coming through the ceiling from an upstairs brothel. One place is reportedly where George Washington played poker and had his stomping grounds.

Who and what to believe? It's all in good fun and there's likely some sliver of truth mixed in with all the rumors.

The War Between the States may have been over long ago, but there's an ongoing war between Annapolis and **Eastport**—a community on the other side of Spa Creek Bridge from Annapolis and technically in a different zip code. Years ago, Annapolis was going to close down the bridge for repairs. Residents of Eastport were upset, to say the least. The boating community in Eastport decided to secede from the Annapolis community as MRE—the Maritime Republic of Eastport.

Secession didn't actually occur, but now there's an annual burning of socks in the spring and donning of socks in the fall, for no socks are worn with boat shoes. The two "towns" have a tug-of-war in the fall, with the winner having bragging rights for a year. Ropes are stretched across the creek and the tug-of-war takes place between the Annapolis Police Depart-

The flag of the Maritime Republic of Eastport flies in a residential neighborhood.

ment and the Fire Department of Eastport. It's all in fun, a chance to drink beer and consume food to benefit charity. There are several different pulls, including one for ladies and one for men over 50. These are comic events, with bartenders from each side pulled straight into the water. Everything is tongue in cheek.

Besides the supposedly longest tug-of-war rope in the world and a constantly friendly competition, both Annapolis and Eastport have numerous restaurants, water taxis, and electric shuttles (glorified electric golf carts but bigger than that) called eCruisers with carbon-free footprints.

If you get a chance, walk by the **Maryland State House**; it dates from the 1770s and it's the oldest state house in continuous legislative use in the United States. In fact, it was the U.S. capitol from November 1783 through August 1784, when the Continental Congress met there.

State laws about slavery and the Underground Railroad were voted on at the State House. There's even a statue of Supreme Court Justice Roger B. Taney on the grounds; Taney ruled on the 1857 Dred Scott case, denying a former slave the right to sue in federal court. Yet there's also a statue of Thurgood Marshall, the first black Supreme Court justice, who was appointed to the court in 1967.

To balance the excitement and energy in Annapolis, consider driving south about an hour and staying overnight at the **Inn at Herrington Harbour.** It's a lovely spot to have a cozy dinner, a restful night's sleep, and breakfast in Friendsville at **Fabulous Brew Café**. The area south of

SAILING TERMS

If you're a novice who plans to hang around sailors, it is helpful to have at least a token amount of jargon under your belt. Here is a lesson from Sailing 101:

There are a few main types of sails; they look different and perform separate functions.

- The *mainsail* powers the boat. Sailors let it out until it begins to flap, and then they bring it in slightly. This allows it to be trimmed for the current course and wind direction. If the course or wind changes, the mainsail will need to be trimmed to adjust to those changes for optimum efficiency and maximum speed.
- The *genoa* is a secondary sail that allows for turning and extra speed. It helps the mainsail stay aerodynamic. Unlike the mainsail, it is not on a boom so the shape is more curved, allowing for faster speed.
- A *jib* is a smaller, genoa-like sail that stays in front of the mast. It is used in high winds.
- A *spinnaker* is a much larger, bag-shaped sail attached to a pole in the front of the boat, and it pulls the boat when the wind pushes it. A *gennaker* acts like a large genoa, whose shape it resembles, in contrast to that of a spinnaker.

Headway means moving forward.

Line is a length of rope with a specified use.

Port is the left-hand side of the boat, looking forward.

Starboard is the right side of the boat, looking forward.

Trimming is adjusting the angle of the sails.

Annapolis is much quieter, filled with working waterfront villages, yet it still retains the sailing and boating aura so compelling in Annapolis.

To get there, take Solomons Island Road (MD 2) south. Make a left onto Friendship Road (MD 261) and drive to the inn, located at the community of Rose Haven on the Chesapeake Bay. After the activity of Annapolis, the drive and the peaceful inn will be appealing.

When you're heading back to the Baltimore area, take Friendship Road to a right on Solomons Island Road (MD 2). From MD 2, take US 50 west to I-97 north to I-695 (Baltimore Beltway).

IN THE AREA

Accommodations

The Inn at Herrington Harbour, 7161 Lakeshore Drive (MD 261), Rose Haven, 20714. Call 410-741-5100 or 800-213-9438. A marina resort that is making every effort to be green. This used to be a motel, so the rooms are small but the renovation was well done. It's not luxurious, but it is comfortable and the food in the restaurant is superb. There's also a small beach, a pool, free continental breakfast, rocking chairs on room porches, cable, and wireless Internet. Web site: www.herringtonharbourinn.com.

Loews Annapolis Hotel, 126 West Street, Annapolis, 21401. Call 410-263-7777. Web site: www.loewsannapolis.com.

Attractions and Recreation

The Annapolis Bookstore, 35 Maryland Avenue, Annapolis, 21401. Call 410-280-2339. Open daily 10:30 AM-6:00 PM; Fri. and Sat. until 9:00 PM. Web site: www.annapolisbookstore.com.

Annapolis Maritime Museum, 723 Second Street, Eastport, 21403. Call 410-295-0104. Small but excellent museum with fascinating information on the importance of oysters to the waters of the bay. Offers seasonal tours to Thomas Point Shoal Lighthouse. Web site: www.amaritime.org.

Banneker-Douglass Museum, 84 Franklin Street, Annapolis, 21401. Call 410-216-6180. This museum is named after two prominent black

Marylanders—Benjamin Banneker and Frederick Douglass. Banneker was born free and was a self-educated surveyor; Douglass was born enslaved and escaped to freedom via the Underground Railroad in 1838. The museum covers African American history in Maryland from 1630. Open Tues. through Sat. Web site: www.bdmuseum.com.

The Cape Exchange, 1348 Cape St. Claire Road, Annapolis, 21409. Call 410-757-6000. For great local lore and some interesting potential purchases, part consignment and part retail store, with interesting jewelry, antiques, and vintage ('40s and '50s) items. From US 50, take exit 29B to Cape St. Claire Road less than 1 mile; in shopping center on left. Closed Sun. Web site: www.thecapeexchange.com.

Charles Carroll House, 107 Duke of Gloucester, Annapolis, 21401. Call 410-269-1737. Former home of the only Catholic signer of the Declaration of Independence, this partially restored house has terraced gardens overlooking Spa Creek. Open Sun. only. Web site: www.charlescarroll house.com.

eCruisers, P.O. Box 5647, Annapolis, 21401. Call 443-497-4769 or 443-481-2422 (dispatch). You can park your car and walk around town; or you can take a ride on an eCruiser—an electric vehicle with service paid for by local merchants, so it's free to visitors (except for optional tips). Web site: www.ecruisersllc.com.

Kunta Kinte-Alex Haley Memorial, City Dock, Annapolis, 21401. Call 410-841-6920. Alex Haley, the author of *Roots,* is immortalized in a statue here. Haley's ancestor, Kunta Kinte, arrived here in 1767 aboard an African slave ship. Web site: www.kintehaley.org.

Maryland State House, 100 State Circle, Annapolis, 21401. Call 410-974-3400. For a short time, this state house was the nation's capitol. Open daily; visitors need a photo ID for entry.

Navy-Marine Corps Memorial Stadium, 550 Taylor Avenue, Annapolis, 21401. For football games, an entertaining spectator sports option. Web site: www.navysports.com.

Sandy Point State Park, 1100 East College Parkway, Annapolis, 21401. Call 410-974-2149. Open year-round with hiking, fishing, picnicking, and sunbathing on public beaches. Located off US 50 at exit 32. Web site: www.dnr.state.md.us/publiclands/southern/sandypoint.html.

Thurgood Marshall Memorial, Lawyers' Mall, Maryland State House, Annapolis, 21401. Call 410-974-3400. Statue of the first African American Supreme Court justice, who was from Maryland.

United States Naval Academy, Armel-Leftwich Visitor Center, 52 King George Street (inside Gate 1), Annapolis, 21402. Call 410-293-8687. Tours of the Yard are available, either escorted or self-guided. The beaux-arts style architecture throughout the campus is beautiful. If you're hungry, the Drydock is a typical college-type cafeteria open to the public on the lower level of Dahlgren Hall. Visitors over 16 must show a photo ID. Web site: www.navyonline.com.

Watermark Tour, Charters, Cruises, City Dock, Annapolis, 21401. Call 410-268-7601 or 410-263-0033 (dispatch). This company provides charters, cruises, and walking tours. Perhaps the best is the simplest—their water taxi, which shuttles visitors and residents across the waterfront from City Dock to Back Creek and beyond. Water taxi operates Apr. through Oct. Web site: www.watermarkcharters.com.

William Paca House and Garden, 186 Prince George Street, Annapolis, 21401. Call 410-267-7619. This is one of those historic houses where it's best for everyone but the most ardent to skip the rather tiresome house tour and just take in the appealing gardens. Paca was a signer of the Declaration of Independence; the house is a restored Georgian mansion with period furniture. Open daily; limited Jan. through Mar. Web site: www .annapolis.org.

Dining/Drinks

b.b. Bistro, 112 Annapolis Street, Annapolis, 21401. Call 410-990-4646. Coffee shop and bistro located in West Annapolis. Web site: www.bbbistro annapolis.com.

Boatyard Bar and Grill, 400 Fourth Street, Annapolis, 21403. Call 410-216-6206. Popular hangout for sailors. Web site: www.boatyardbarand grill.com.

Ego Alley at City Dock, where those with big boats come to show them off

Carrol's Creek Waterfront Restaurant, 410 Severn Avenue, Annapolis, 21403. Call 410-263-8102. Superb cream of crab soup. Excellent fish and seafood entrees. Perfect view of historic Annapolis and recreational boats on Spa Creek to the left and the Severn River to the right. Web site: www.carrolscreek.com.

Chick & Ruth's Delly, 165 Main Street, Annapolis, 21401. Call 410-269-6737. At this political hangout, everyone (customers and employees alike) recites the Pledge of Allegiance at 8:30 AM weekdays and 9:30 AM weekends. Dishes are named after state politicians. Open daily for breakfast, lunch, and dinner. Web site: www.chickandruths.com.

Fabulous Brew Café, 11 West Friendship Road, Friendship, 20758. Call 410-286-8799. A wonderful find in an obscure location. What's not to like at this café where the service is, of course, friendly and the food is good? Open for three meals daily except for dinner Mon., and only early dinner Wed. And they'll do egg whites for the egg dishes and omelets! Close to Herrington Harbour, so you'll want to stop here at least once.

Galway Bay, 63 Maryland Avenue, Annapolis, 21401. Call 410-263-8333. Irish pub. Web site: www.galwaybayannapolis.com.

Jimmy Cantler's Riverside Inn, 458 Forest Beach Road, Annapolis, 21401. Call 410-757-1311. On Mill Creek, this is the perfect setting for eating steamed hardshell crabs. Lots of atmosphere in this crab house, but stick to eating here in the summer. It can get deserted at other times, which makes it much less fun. Open daily for lunch and dinner. Web site: www.cantlers.com.

Leeward Market Café & Pizzeria, 601 Second Street, Annapolis, 21403. Call 443-837-6122. Breakfast, deli sandwiches, pizza, and coffee drinks. Open daily for breakfast and lunch.

Mango's Bar & Grill, 7153 Lake Shore Drive, Herrington Harbour South, Rose Haven, 20714. Call 410-257-0095. Web site: www.mangoson thebay.com.

Rams Head On Stage, 33 West Street, Annapolis, 21401. Call 410-263-1167. For those who like food and drink with their entertainment, this is a good place. Concerts and comedy acts in a comfortable venue for 21 and over only. Web site: www.ramsheadtavern.com/annapolis/onstage .html.

Sam's on the Waterfront, 2020 Chesapeake Harbour Drive East, Annapolis, 21403. Call 410-263-3600. Absolutely superb food. Web site: www.samsonthewaterfront.com.

More Contacts

Annapolis & Anne Arundel County Conference & Visitors Bureau, 26 West Street, Annapolis, 21401. Call 410-280-0445 or 888-302-2852. Web site: www.visitannapolis.org.

Four Rivers: The Heritage Area of Annapolis, London Town & South County, The Arundel Center, 44 Calvert Street, Annapolis, 21401. Call 410-222-1805. Web site: www.fourriversheritage.org.

The grounds of Rocky Gap Lodge in Flintstone

CHAPTER

9

Historic Western Maryland

Trekking through the Scenic Mountains

Estimated length: 250 miles
Estimated time: Weekend

Getting there: From Frederick, take a combination of I-70 west and US 40 west to the Hancock area, where you'll connect to I-68 west (I-70 turns north into Pennsylvania). Follow I-68 west through the town of Flintstone and on to Cumberland. Take exit 43C and follow the signs to Canal Place and the Western Maryland Railway Station in downtown Cumberland.

Highlights: Deep Creek Lake. Railroad town of Oakland. Cumberland and the C&O Canal. Western Maryland Scenic Railroad. Town of Frostburg. Towns named Accident and Flintstone.

Mountains excite people. They are impressive, beautiful, and cool in summer, and they provide opportunities for recreational activities year round. Western Maryland is mountainous and beautiful; it has the added benefit of being close to Baltimore, Washington, DC, and Pittsburgh. That makes it a great destination for a weekend trip.

There's the added benefit of traveling on at least part of the Old National Road, which is now US 40. This historic road was originally an Indian trail called Nemacolin's Path that crossed the Appalachian Mountains. For many years, the road was the main route connecting the East with the Ohio Valley. In the 1800s, the federal government paid to rebuild the road from Cumberland, Maryland, to Wheeling, West Virginia, making it the first national highway.

When you're passing through **Hancock** and connecting to I-68, you are in the skinniest or narrowest point in the state, where Maryland is only about 4 miles wide from Pennsylvania on the north to West Virginia on the south. You'll pass by the tiny town of **Flintstone,** which is generally great fun for kids who want to say they've "been to Flintstone."

If you're looking for a great place to stay overnight, **Rocky Gap Lodge and Golf Resort** in Flintstone is a comfortable mountain resort that makes visiting Rocky Gap State Park an upscale experience with the resort's spa, golf course, lake, and beach. The lake is a small man-made one right by the lodge, and the birds and small animals outside are a never-ending source of amusement for children, who often stand fascinated by the window of the restaurant while the adults finish dining.

If you're continuing on, you'll come to **Cumberland**, where, as with many places in Maryland, transportation dictated the growth of the town. In this case, the railroad and the C&O Canal each played a part. By stopping at the Western Maryland Railway Station, you can get a glimpse into both for there's the marvelous **C&O Canal Visitor Center** on the ground floor of the train station, where you can gain an appreciation for the importance of the waterway in the 1800s.

If you time it right, you can also take a round-trip excursion train with a steam locomotive from the station in Cumberland to **Frostburg,** about an hour away. It's an eye-opening ride, with the slow speed and thick black smoke, and yet there's an indescribable allure of the past romance of the rails.

You can also take a look at Cumberland from the endpoint of the **C&O Canal towpath.** The canal stretched from Georgetown in Washington, DC, to its endpoint in Cumberland—a distance of 184.5 miles. The shipping waterway is no longer in use, in fact some of it is dry, but the towpath, along which mules pulled the canal barges, is now managed by the National Park Service and is a favorite of hikers and bikers.

Cumberland is a great jumping-off point for bikers, as you can take the C&O Canal towpath south toward Washington. Cumberland is also where the Great Allegheny Passage begins—a 150-mile trail that leads north toward Pittsburgh.

While in Cumberland, you can wander over to the tiny commercial district where there are a few shops and restaurants. Or you can get back on I-68 and keep going west.

All along these mountain roads, keep alert for wildlife. People share these roads with deer, bear, and other critters; you wouldn't want to hit any of them. The road signs warn about deer crossings; they also warn about bears. That tells you that this is an isolated mountainous area—perhaps it's not how we generally think about Western Maryland, but this is one perspective.

Next you'll come to Frostburg, where you should venture into town and walk along Main Street. If you're hungry, stop in at the **Princess Restaurant** and have one of their sandwiches made on delicious grilled toast. Or stop at the **Hen House Restaurant** outside of town on the National Pike (Alternate US 40); their homemade chicken noodle soup and barbecued ribs are delicious.

Back on the way west, get off I-68 at exit 19 and onto Alternate US 40 a half-mile east of **Grantsville** to see the 1813 **Casselman River Bridge,** which was the country's longest single-span stone arch bridge when it was built. It's no longer in use; there's a newer bridge right next to it. But the old one is pretty and interesting.

At this point, you're very close to **Savage River State Forest.** Inside the state forest, on private property, is **Savage River Lodge.** To get there from I-68, take exit 29 (Finzel Road), turn south onto Beall School Road (MD 546), and follow the signs to the lodge.

The lodge is delightfully remote and appealing, and guests are treated to private log cabins with many touches of luxury, including plush robes, king- or queen-sized beds, and an excellent on-site restaurant that will deliver to your cabin if you really want privacy. No TVs here, but there are many opportunities for hiking, cross-country skiing, fly-fishing (rivers and streams), biking, white-water rafting, and more. It seems as if you're in another country, not just in another part of the state.

The driveway is 1 mile off the secondary road, which goes from paved to gravel, and it all has a jungle lodge feel. You can stay here or just arrange to have a meal at the restaurant if you're not sure how you'll feel in the total isolation created by the surrounding thick, dark forest. It is an experience. If you have a vivid imagination, you might want to give it another thought. But it is beautiful, and remote...with little if no cell connectivity. You might want to call before you go—to make reservations for a stay or a meal and to get more detailed directions.

Back on the road, continue driving on I-68 to exit 14 and take US 219 south. You're heading toward the town of **McHenry** and **Deep Creek Lake.**

There are many rivers and creeks in Maryland, but no natural lakes. The largest man-made lake in the state is Deep Creek Lake, which was created by building a dam on Deep Creek. The creek is a tributary of the **Youghiogheny River** (pronounced "Yock-eh-gain-y" and abbreviated by locals as "the Yock" to rhyme with "jock").

In constructing the dam on the "Yock," the original intent was to harness energy. In the process, and as an added bonus, those responsible created a marvelous recreational destination in the heart of the Western Maryland mountains. This is part of the Allegheny mountain range and between the mountain vistas and the lake, there is real beauty.

Deep Creek Lake is 12 miles long with a 65-mile shoreline. It has become somewhat crowded in recent years. It is still extremely pretty, and it provides a major year-round vacation and weekend destination. The winter recreation is intense: downhill skiing, snowboarding, snowshoeing, cross-country skiing, sledding, snowmobiling, ice fishing. Summer recreation is just as plentiful: boating, waterskiing, canoeing, sailing, swimming, fly-fishing, hiking, biking, birding, white-water rafting.

Then there are those who sit and read a book. Or have the most delicious homemade ice cream imaginable at the **Lakeside Creamery.** Or play poker, or go to the movies, and just plain relax.

You might consider driving a bit farther to stay in nearby **Oakland**, which is close to the lake but smaller and quieter; from Oakland, you can drive the 12 or so miles to Deep Creek whenever you want (winter weather permitting).

Oakland is an interesting destination in and of itself. There's an antiques mall with an old-fashioned soda fountain in back and a picturesque old historic B&O Railroad station dating from 1884. Although the town of Oakland is the county seat, it is a sparsely populated, mountainous county, so that equates to a small town—a great small town, but a small one nonetheless.

There's a wonderful bed & breakfast in Oakland where you can stay— **The Oak and Apple Bed & Breakfast.** The hosts are delightful and the B&B is located in a beautiful house where you'll likely feel right at home. To get there, stay on US 219 (Garrett Highway) past Deep Creek Lake for another 12 miles.

Once in Oakland, turn right onto East Crook Street and go one block to Second Street. The B&B is on the corner, with parking at the end of a circular drive. Rooms are comfortable and come with private baths, and there's a TV room for guests to use so they don't have to sit in their bedrooms. The hosts admit to their dual loves of "cooking and chatting," so breakfast is both delicious and conversationally stimulating.

THE WAY WEST

When our country was founded in the late 18th century, its western boundary was marked by the Appalachian Mountains. President George Washington proposed a series of national roads to help quell Native American uprisings, repel foreign invasions, and put down potential tax rebellions.

In 1806, when Thomas Jefferson was president, Congress approved construction of the National Road, the country's first federally funded highway. When constructed, the National Road began in Cumberland and went west to Wheeling and beyond, improving and extending the previously existing road (now US 40). The road east from Cumberland to Baltimore was privately financed by banks, which underwrote bonds and sold them to investors.

US 40 both parallels and intersects I-68, which was built later across the western part of Maryland. In some places, 40 also splits into two parts—the Scenic or Alternate 40 and the regular 40. It may seem confusing, but once you're traveling you'll see that you can go back and forth from one to the other with relative ease. On US 40 and US 40 A, you may see some of the original markers indicating the distance from Baltimore in one direction and the distance forward to various other cities and towns. You can make a game of looking for the markers, or just relax and enjoy the scenery.

The National Road was vital to the growth of the nation. It was actually a turnpike road and there's still a tollhouse in **LaVale,** between Cumberland and Frostburg, which you can stop and tour. Even if you happen upon the tollhouse when it's closed, stop anyway and look around, for there are signs posted that tell about the cost for transportation in the days when everyone was in a wagon or on horseback.

Travelers should be aware that the historic National Road is comprised of several roads that have varying names in different places. These include Baltimore National Pike, US 40, Scenic US 40, Alternate US 40 or US 40A, I-68, MD 144, and Frederick Road. Some of these roads are located east of Cumberland and thus are not technically part of the federally funded portion that started in Cumberland and went west; however, all the roads mentioned here are usually considered part of the National Road.

Another mode of transportation, via a constructed waterway, was the C&O Canal, which, in its time, competed with the Baltimore and Ohio (B&O) Railroad for the transport of freight. Perhaps it was the cumbersome nature of the lock system, the potential for flooding, or the superiority of the railroad that led to the ultimate failure of the canal.

Cumberland is located at the last completed spot on the C&O Canal route. Construction of the canal started on July 4, 1828. It was originally intended to be 360 miles long but never reached that length, as it failed to meet expectations, especially after the B&O Railroad spurted ahead in the competition for transportation of people and goods.

Flooding in 1924 destroyed much of the canal and it was never reopened. It was turned over to the federal government in 1938 and proclaimed a national monument in 1961 and a national historic park in 1971. The towpath is now used for recreational purposes.

Oakland grew after the B&O came into the Western Maryland region in 1851, becoming a railroad town and a headquarters for the empire of John W. Garrett, who was president of the B&O at that time. In 1872, the Maryland assembly decided to name what is now Garrett County in honor of him.

Once the B&O laid train tracks and built its stations, and before air-conditioning became a fact of life, residents of Baltimore and Washington, DC, sought to spend summertime in the mountains around Oakland. To accommodate this desire and to encourage train travel, the B&O built the Oakland Hotel in 1875 as well as the nearby Deer Park Hotel by the natural spring there.

These and other grand hotels are long gone, lost to either fire or razing. Summer tourism declined in the early 1900s after cars surpassed train travel as the preferred mode of transportation. Tourism returned in earnest in the late 1920s once the dam was built on Deep Creek, forming the lake.

Besides tourism, the railroad was important for transporting lumber and coal out of the area, and for putting money into local businesses.

There is much to see in Oakland, including the **B&O train station,** the **Garrett County Historical Society Museum,** the **Book Mark'et & Antique Mezzanine, Englander's Antiques & Grill**—an antiques mall with different dealers plus an old-fashioned soda fountain in back—and the **Sam Pearce** hand-blown glass factory outside of town.

There's also **St. Matthew's Episcopal Church,** which was originally

A steam locomotive ride through the countryside on the Western Maryland Scenic Railroad

Garrett Memorial Presbyterian Church, completed in 1869 with money from John Garrett as a memorial to his brother, Henry. Five U.S. presidents worshiped here when they came to Oakland.

While you're staying in Oakland, if you want to take a pretty drive, go north on US 219 and then connect onto MD 42 north into **Friendsville,** which is a kayaker's dream, for the Youghiogheny River runs through town. White-water rafting is also popular. Back in the day, settlers lived peacefully with Native Americans in this area. Friendsville was prosperous in the 1820s from the iron industry. Coal mining later brought the railroad, and lumbering brought more affluence. But when the Youghiogheny Dam was built, the railroad left, as did many people; when the railroad withdrew, the prosperity went with it. Now Friendsville is a sleepy community with paddle sports and other water recreation.

Between Deep Creek Lake and I-68 is the town of **Accident,** which you'll pass through on US 219, as if by…accident! When Western Maryland was opened for settlement by Lord Baltimore in 1774, two surveyors were surveying the land. The story is that they both selected the same tract of land to survey, as if "by accident." The name stuck.

Deer Park is close to Oakland along MD 135 and it is the source of bottled water for lots of people. It's also one of the places where the B&O brought wealthy people in the summer to enjoy the cool mountain climate and a luxurious resort hotel. The hotel in Deer Park opened in 1872 and eventually had three hundred rooms. Unfortunately, the hotel is long gone, but Deer Park Natural Spring Water still comes from this town.

When your weekend is over and you've crammed in as many of these places as possible, or as few if you just want to breathe in the clear mountain air and relax, head north on US 219 (past Accident) to I-68 (to Hancock) to I-70 east to Baltimore. Until the next time the lure of the mountains calls….

IN THE AREA

Accommodations

Casselman Inn, 113 East Main Street (Alternate US 40), Grantsville, 21536. Call 301-895-5055 (hotel) or 301-895-5266 (restaurant). Located on the Historic National Road, this inn was built to cater to stagecoaches and covered wagons. Now there are inexpensive rooms for rent. It's interesting, but does not have the most comfortable accommodations. However, there is a pleasant restaurant downstairs where you can get a decent breakfast. Take I-68, exit 19. Web site: www.thecasselman.com.

The Oak and Apple Bed & Breakfast, 208 North Second Street, Oakland, 21550. Call 301-334-9265. A lovely large house, with hosts John and Brenda Rathgeb. There's a big porch with rocking chairs, satellite TV, and wireless Internet, and in winter great opportunities for snowshoeing and cross-country skiing. Plus Oakland is a delightful town. Web site: www .oakandappleinn.com.

Rocky Gap Lodge & Golf Resort, 16701 Lakeview Road NE, Flintstone, 21530. Call 301-784-8400 or 800-724-0828. Comfortable mountain lodge in Rocky Gap State Park. Spa, golf course, and lakeside beach and boating. Right on I-68. Web site: www.rockygapresort.com.

Savage River Lodge, 1600 Mt. Aetna Road, Frostburg, 21532. Call 301-689-3200. On private land in the midst of Savage River State Forest—the state's largest forest. From here you can go white-water rafting, fly-fishing, and biking on the Great Allegheny Passage. Web site: www.savage riverlodge.com.

Wisp Resort & Hotel, 296 Marsh Hill Road, McHenry, 21541. Call 301-387-5581, 301-387-4911, or 800-462-9477. Downhill skiing in winter. Web site: www.wispresort.com.

Attractions and Recreation

B&O Railroad Station, Oakland. With 1884 Queen Anne–style architecture, the station is a centerpiece for this beautiful little town.

Barn Quilt Association of Garrett County, 809 Memorial Drive, Oakland, 21550. Call 877-577-2276. There's a movement afoot to paint quilt designs and mount them onto barns. It's taking hold in many places, including the mountains of Western Maryland. You can find a trail of these barn quilts on this association's Web site: www.garrettbarnquilts.org.

Book Mark'et & Antique Mezzanine, 111 South Second Street, Oakland, 21550. Call 301-334-8778. A charming, well-stocked, independent bookstore. Open daily.

Casselman River Bridge, 10240 National Pike, Grantsville, 21536. Call 301-895-5453. Built in 1813, it was the longest single-span stone arch bridge in the country. No longer in use but it's worth a look. Web site: www.dnr.state.md.us/publiclands/western/casselman.asp.

C&O Canal Visitor Center and Museum, 13 Canal Street, Cumberland, 21502. Call 301-722-8226. Located on the ground level of the Western Maryland Railway Station. Towpath is nearby. Open daily. Web site: www.nps.gov/choh.

Alpacas at the River's Edge Farm

Cumberland Trail Connection, Canal Place, Cumberland, 21502. Call 301-777-8724. Bicycle sales and rentals, plus related equipment.

Deep Creek Lake State Park, 898 State Park Road, Swanton, 21561. Call 301-387-5563. Web site: www.dnr.state.md.us/publiclands/western/deep creeklake.html.

Englander's Antiques & Grill, 205 East Alder Street, Oakland, 21550. Call 301-533-0000. Antiques mall plus old-fashioned soda fountain in back.

Garrett Eight Cinemas, 19741 Garrett Highway, Oakland, 21550. Call 301-387-2500. Web site: www.garrett8cinema.com.

Garrett County Historical Society Museum, 107 South Second Street, Oakland, 21550. Call 301-334-3226. Web site: www.deepcreeklake.com /gchs/index.asp.

LaVale Toll House, 14302 National Highway, LaVale, 21502. Call 301-777-5132.

Mountain City Traditional Arts, 25 East Main Street, Frostburg, 21532. Call 301-687-8040. Exhibits and objects for sale that feature Appalachian arts and crafts. Web site: www.mountaincitytradearts.com.

Queen City Transportation Museum, 210 South Centre Street, Cumberland, 21502. Call 301-777-1776. Historic vehicles range from an 1840 Conestoga wagon to a 1920 Buick Roadster. Days and hours vary. Web site: www.queencitytransportationmuseum.com.

River's Edge Alpaca Farm, 20020 Potomac Overlook, Oldtown, 21555. Call 301-478-5424. Working alpaca farm where you can pet the sweet-looking animals. Located on a true backroad on the edge of Maryland across the Potomac River from West Virginia. Call first for directions and to arrange a visit. Web site: www.barbsalpacas.com.

Simon Pearce, 265 Glass Drive, Mountain Lake Park, 21550. Call 301-334-5277 or 800-774-5277. This is an interesting hand-blown glass factory where you can take a self-guided tour to watch the artisans create pitchers, vases, and more. Learn about the process of glassblowing from composition and melting to blowing, shaping, and finishing. Shop in the attached retail store. Open daily. Web site: www.simonpearce.com.

Spruce Forest Artisan Village, 177 Casselman Road, Grantsville, 21536. Call 301-895-3332. Open year-round; free admission. Web site: www.spruceforest.org.

Thrasher Carriage Museum, 19 Depot Street, Frostburg, 21532. Call 301-689-3380. Web site:thethrashercarriagemuseum.com.

Western Maryland Scenic Railroad, 13 Canal Street, Cumberland, 21502. Call 301-759-4400 or 800-872-4650. Take a 32-mile round-trip ride on a steam or diesel locomotive excursion through the Allegheny Mountains from Cumberland to Frostburg and back. Evening Murder Mystery excursions too. Varied schedule May through Dec. Web site: www.wmsr.com.

Dining/Drinks

Black Bear Restaurant & Tavern, 102 Fort Drive, McHenry, 21541. Call 301-387-6800. Bar food, pool, sports on multiple TVs. Web site: www.blackbeartavern.com.

City Lights, 59 Baltimore Street, Cumberland, 21502. Call 301-722-9800. Sandwiches, salads, and wraps for lunch plus beef and crab dishes for dinner. Open Mon. through Sat. for lunch and Tues. through Sat. for dinner. Web site: www.citylightsamericangrill.com.

Cornish Manor, 830 Memorial Drive, Oakland, 21550. Call 301-334-6499. Victorian mansion turned restaurant. Open daily for lunch and dinner. Web site:www.cornishmanor.net.

The Deer Park Inn, 65 Hotel Road, Deer Park, 21550. Call 301-334-2308. Web site: www.deerparkinn.com.

Hen House Restaurant, 18072 National Pike, Frostburg, 21532. Call 301-689-5001. Excellent food, large quantities, and a comfortable country setting. The chicken noodle soup is good, the barbecued ribs are superb. You can also get steamed hardshell crabs in season, and sit inside or out on the patio. When you're nearby, make time to have a meal here. Web site: www.henhouserestaurant.com.

JG's Pub, 18553 Garrett Highway, Oakland, 21550. Call 301-387-6369. Food, drink, pool, and keno.

Lakeside Creamery, 20282 Garrett Highway, Oakland, 21550. Call 301-387-2580. Open during the summer on the shore of Deep Creek Lake. Fantastic ice cream sold by the ounce rather than the scoop; you can mix flavors or keep it simple. Web site: www.lakesidecreamery.com.

Mountain City Coffeehouse & Creamery, 60 East Main Street, Frostburg, 21532. Call 301-687-0808. Coffee, baked goods, soups, salads, and ice cream from Lakeside Creamery in Deep Creek. Closed Mon. Web site: www.mtncitycoffeehouse.com.

Penn Alps Restaurant & Craft Shop, 125 Casselman Road, Grantsville, 21536. Call 301-895-5985. Open weekdays. Web site: www.pennalps.com.

Princess Restaurant, 12 West Main Street, Frostburg, 21532. Call 301-689-1680. Old-fashioned counter and booth restaurant. Great sandwiches made with delicious grilled toast. Closed Sun. Web site: www.princess restaurant.com.

Santa Fe Grille, 75 Visitors Center Drive, McHenry, 21541. Call 301-387-2182. Excellent Southwestern fare. Web site: www.dclsantafe.com.

Other Contacts

Allegany County Department of Tourism, 13 Canal Place, Cumberland, 21502. Call 301-777-5134 or 800-425-2067. Web site: www.md mountainside.com.

Downtown Cumberland Business Association, 1 Washington Street, Cumberland, 21502. Call 301-722-5500. Web site: www.visitcumberland .org.

Garrett County Chamber of Commerce, 15 Visitors Center Drive, McHenry, 21541. Call 301-387-4386. Web site: www.visitdeepcreek.com.

Town of Oakland, 15 South Third Street, Oakland, 21550. Call 301-334-2691. Web site: www.oaklandmd.com.

MASON AND DIXON LINE
105TH MILE STONE

500 FEET BEYOND THIS POINT, ON PRIVATE
PROPERTY, THIS STONE IS LOCATED.
IT BEARS THE COAT OF ARMS
OF LORD BALTIMORE AND WILLIAM PENN.
THE 104TH MILE STONE AND 103RD MILE
STONE BEAR THE LETTERS M AND P
MARYLAND-PENNSYLVANIA AND ARE
LOCATED ALONG THE MARYLAND EDGE
OF THIS ROAD.
MARYLAND HISTORICAL TRUST
MARYLAND STATE HIGHWAY ADMINISTRATION

Sign pointing to a Mason-Dixon Line crown stone

CHAPTER 10

Antietam Battlefield and the Mason-Dixon Line

Retracing the Steps of the War between the North and South

Estimated length: 160 miles
Estimated time: Overnight

Getting there: From Baltimore, take I-70 west past Frederick to exit 49 (Braddock Heights). At the light, go left (north) onto Alternate US 40. Stay on 40A for 11 miles until you reach the light in Boonsboro. The Inn BoonsBoro is on the right, just past the light on the square; parking is around back.

Highlights: Antietam National Battlefield—the state's largest Civil War battlefield and site of the bloodiest day of the four-year war. The sleepy town of Boonsboro. A lovely inn owned by romance novelist Nora Roberts. Peaceful scenery. Mason-Dixon Line markers near the Pennsylvania border.

This trip entails a great deal of drama—both the Civil War kind and the literary kind. You'll be staying in the picturesque and historic town of **Boonsboro** at an inn designed and owned by best-selling romance novelist Nora Roberts, close to one of the most important battlefields of the War Between the States.

Boonsboro is a charming, somewhat sleepy little town along the Historic National Road about midway between Frederick and Hagerstown.

Some of the high points in a visit to this town are attributable to Nora Roberts, whose best-selling romance novels are legendary in bookstores and airport newsstands. She owns the **Inn BoonsBoro,** and her husband owns the café and bookstore **Turn the Page Bookstore Café,** which is across the street. Together they recently opened a gift shop, **Gifts Inn BoonsBoro,** next door to the bookstore.

Boonsboro dates back to 1792, when William Boone and his brother, George, established the town. (William and George Boone were relatives of the famous Daniel Boone.) In 1810, the road from Baltimore to Boonsboro was completed, bringing prosperity to the area, for the National Road passed through the town as Main Street. The 1820 census showed the population consisted of 395 whites, 7 free blacks, and 26 slaves. Later the town figured prominently in the Civil War, with nearby battles at South Mountain and Antietam, and with armies on both sides passing through on several occasions. Minor battles were even fought in the heart of Boonsboro, right on Main Street.

The town is now a bedroom community for both Frederick and Hagerstown, with a surprisingly large amount of traffic, for it is located at the intersection of US 40A and MD 34. Watch out for the traffic as you cross busy Main Street. Yet traffic aside, Boonsboro retains its small-town flavor.

For a base of operations, you won't find anything more appealing than Inn BoonsBoro. The rooms are decorated to highlight the owner's love of romantic couples in literature, including the Nick and Nora room from Dashiell Hammett's *The Thin Man,* the Jane and Rochester room from Charlotte Brontë's *Jane Eyre,* and the Elizabeth and Darcy room from Jane Austen's *Pride and Prejudice.* The inn is a wonderful place to stay with delightful private rooms plus shared spaces that include a library, porches, dining room, and den. Guests also have access to the refrigerator in the kitchen and can help themselves to bottled water and other drinks.

Make sure to walk around town a bit and visit the bookstore to get a good idea just how prolific Nora Roberts is as an author. In addition to carrying all of the books by Nora Roberts, the store has an interesting selection of other books too. Or you can watch the town from the prime vantage point of the inn's second-floor front porch.

Boonsboro seems so peaceful that it is hard to imagine it was so much a part of several brutal Civil War battles. In fact, General Robert E. Lee actually used this now quiet little town for his headquarters during the Battle of South Mountain.

Maryland was, after all, one of the Civil War–era border states, and it certainly exemplified the "house divided" concept and the "brother fighting brother and neighbor fighting neighbor" aspect of the war. Part of the state was sympathetic to the anti-slavery cause and the other was entrenched in slave plantations as a way of life. There were other volatile issues of the day too, and all this discord was propelling the state into conflict over which side to support.

With Washington, DC, bordering Maryland (DC was formed from land that once belonged to Maryland), the state's precarious political stance was of grave concern to President Abraham Lincoln and his military leaders. Attempts by the Confederates to take over the capital city would be devastating to morale, not to mention to the running of the war and the government.

Across the Potomac on the other side of the nation's capital, Virginia was mostly Confederate territory (with small pockets loyal to the Union); and the headquarters of the Confederacy was not far away in Richmond. If Maryland fell to the South, then Washington would have been surrounded by the enemy. President Abraham Lincoln and his generals were determined to prevent this from happening. Meanwhile, Confederate president Jefferson Davis and General Robert E. Lee sought to do just that—cut Washington off from the Northern states by invading Maryland.

Antietam National Battlefield, in the mountainous western part of Maryland, was the site of the bloodiest one-day battle of the Civil War on September 17, 1862. The battle was somewhat of a draw, but it resulted in the North managing to thwart the Southern advance into Maryland. Both sides suffered tremendous loss of life and limb, yet when one visits the battlefield with its scattered cannons and monuments, the peaceful countryside seems like anything but a battlefield.

To get to Antietam, retrieve your car from the parking lot in back of the inn and turn right onto Shepherdstown Pike Road (MD 34) going west. Antietam National Battlefield is located in Sharpsburg, which is only 8 miles away. In between, you pass the town of **Keedysville.** If you're hungry, stop at the **Red Byrd Restaurant** for something to eat. Red Byrd is not fancy, but you can get an inexpensive and decent meal. From Keedysville, continue on MD 34 a few more miles and turn right onto Sharpsburg Pike

(MD 65). You are on the Antietam National Battlefield grounds and the first stop is at the Visitor Center to pick up a printed guide to the national park.

Antietam is to Maryland what Gettysburg is to Pennsylvania—a vitally important battlefield that epitomizes the conflict between proponents of states' rights and champions of strong federal power in the mid-1800s. Slavery was one issue, but it was not the only one. Self-determination, taxes, and big brother government before anyone heard of big brother—all played a part.

There's certainly an ongoing fascination with Civil War battles and the lands where they took place. Antietam embodies the essence of the war as well as any other. Plus this battlefield was particularly chaotic. Weapons often misfired and troops went in the wrong direction, creating confusion and leading to unnecessary death and injury.

At the Visitor Center, get a copy of the National Park Service brochure guide with a map of the battlefield that includes numbered stops along a driving tour. By skipping the ranger-led caravan and doing the tour yourself, you have the freedom to proceed at your own pace. It's easy to get a feel for the terrain by using your imagination to picture the fighting done at close range among the cornfields. To assist you, this battlefield is kept as true to the 1862 battle conditions as possible.

Follow the Visitor Center guide as well as the small blue arrowed signs. You need not go to all the stops, but be sure to see the **Cornfield,** where much of the morning phase of battle occurred (stop 4). And you don't want to miss **Bloody Lane,** otherwise known as the "Sunken Road" (stops 7 and 8). Soldiers under the command of Robert E. Lee hid in the worn wagon road that was the Sunken Road, while Union troops under General McClellan crested over the edge and were taken down as the Confederates picked them off. Finally the Union soldiers were able to outflank the Southern soldiers.

Be sure to also visit **Burnside Bridge,** which the Union soldiers had trouble crossing due to Confederate sharpshooters. Located on Sharpsburg Pike (MD 65), the bridge is one of many in this part of the state; built in 1834, it is representative of the stone arched bridges built prior to the Civil War. During the war itself, both sides crossed and recrossed the bridges numerous times. This particular bridge was nicknamed for Union general Ambrose Burnside; he was the general charged with crossing the creek and taking the bridge despite stubborn and well-entrenched Confederate defenses.

HOW TO MAKE THE MOST OF YOUR VISIT TO ANTIETAM

Don't worry if you're not a Civil War buff and don't try to remember too many details on any one visit. You can still appreciate the military action that forced the Confederates to withdraw.

Absorb the peacefulness of the countryside, for the raging battle is long since past. Focus on terrain that is much the same as it was then, for the battlefield is being preserved to mirror the way it was on that pivotal day in 1862. Recognize that the battle led to Lincoln's issuance of the Emancipation Proclamation, which in effect freed the slaves in states that were rebelling against the Union, and then became law for the entire country.

Notice the Observation Tower at one end of Bloody Lane, which was built by the War Department in 1896 to aid in studying the battlefield.

Look for cannon markers memorializing generals who were killed or mortally

Now-peaceful vista at Antietam with artillery on a high position

wounded during the battle—three from the North and three from the South. These generals are acknowledged with upturned cannon barrels mounted in blocks of stone.

Notice the 96 monuments on the battlefield, which were erected in the 1880s and 1890s. The majority are Union monuments, for the former Confederacy was financially devastated after the war, making it difficult for the South, which suffered the economic repercussions of defeat and Reconstruction, to raise money for building monuments.

Look for the statue dedicated to Clara Barton, who helped found the Red Cross after the Civil War; she worked hard to care for the wounded at this battle.

As in other areas of Maryland, North–South sympathies in this region depended in part upon who was in charge. The locals considered themselves Yankees when the Yankees were there. They considered themselves Confederates when the Confederates were there. One wonders if more Marylanders would have claimed they sided with the Confederacy if the South had won.

After your visit to Antietam, you can enjoy dinner in Boonsboro, either at **Palettie** across the street from the inn or at **Old South Mountain Inn** down the road by going east on Alternate US 40. You'll also want to enjoy the Inn BoonsBoro, so you should plan to go back to your room early, for there is a flat-screen TV and DVD player in each room, with a library of DVDs for your use. The rooms are supremely comfortable, the baths and showers so high tech (the toilets even flush themselves) that you are sure to enjoy the indulgences.

In the morning, breakfast is included with your stay. Or you can go next door and get something to eat at **Icing Bakery.** Then you'll head north toward the Maryland–Pennsylvania border to search out two **Mason-Dixon Line markers.**

The Mason-Dixon Line was the philosophical, but not the actual, demarcation line between Northern and Southern sympathies during the Civil War. While Maryland stayed with the North due in large part to military pressure exerted from Washington, DC, the state was split in terms of its loyalty. The Mason-Dixon Line separated Maryland from Pennsylvania in the north (and Delaware in the east); the line was culturally symbolic, even though the Potomac River turned out to actually separate north from south.

In the 1760s (1763–68 to be exact), two English surveyors—Charles Mason and Jeremiah Dixon—searched and documented the northern bor-

OTHER BATTLES

While Antietam is the largest battlefield in Maryland, two other significant battles took place nearby.

Just three days before the dire events at Antietam, the **Battle of South Mountain** took place on September 14, 1862; it was the first major battle on Northern soil. The battlefield contains **Gathland State Park** near the village of **Burkittsville.** Of interest is the impressive **War Correspondents Memorial Arch.**

At the Battle of South Mountain, fierce fighting took place over possession of three pivotal gaps in the mountains—Crampton's, Turner's, and Fox's Gaps. To see the battle site for yourself, drive on Alternate US 40 east from Boonsboro to MD 67 south and follow the signage.

Monocacy National Battlefield is close by too, and it is another pleasant farmland scene that witnessed an important Civil War battle. When Richmond was under siege in 1864, Confederate general Jubal A. Early tried to invade DC by advancing through Maryland. Although the North was defeated here on July 9, 1864, Union troops still managed to delay the Confederate Army for one day, thus allowing Union military reinforcements to reach the nation's capital. When Early and his troops reached Washington on July 11, they were unable to overpower the additional Union defenses.

A few years earlier, also in a field at Monocacy alongside the railroad tracks and the Monocacy River, Union soldiers discovered Robert E. Lee's Special Orders Number 191 wrapped around three cigars. The discovery of these orders, which constituted military plans for Lee's army, allowed the Union generals insight into the Confederate campaign plan. Lee's plan split up his troops into three parts—one to secure Harpers Ferry, one to guard Turner's Gap near Boonsboro, and the third to enter Hagerstown near the Mason-Dixon Line on the way to invading Pennsylvania. When Union Army generals discovered this plan, they gave chase, forcing Lee to take a stand at Sharpsburg and triggering the Battle of Antietam, with a great loss of life to both sides.

To reach Monocacy, drive on Alternate US 40 east from Boonsboro and take I-70 east toward Frederick. Take exit 54 (Market Street/MD 85). Bear right on MD 85 toward Buckeystown. Turn left (south) at the second light onto Urbana Pike (MD 355). The Visitor Center, 1.5 miles down on your left, is run by the National Park Service.

Close-up of a Mason-Dixon Line crown stone on private property

der of Maryland to settle a bitter land dispute between Maryland and Pennsylvania. (They also surveyed the border between Maryland and what were then the three southernmost counties in Pennsylvania but are now in Delaware.)

The surveyors literally drew a line in the ground, for they placed 1-mile markers across the entire border, which became symbolic of the cultural divide between north and south. Slavery was abolished in Pennsylvania in 1781, but it existed throughout Maryland on various plantations until after the conclusion of the Civil War in 1865. Taxes and land rights were other issues surrounding the Mason-Dixon Line.

Few of the original markers still exist or are easily accessible. After all, centuries have passed and many of the markers are on private lands. The ravages of weather and farm equipment have also taken their toll.

Every fifth stone marker contained the coats of arms of the Maryland and Pennsylvania colonial leaders—for Maryland, the coat of arms for Lord Baltimore (Calvert family) on the south, and for Pennsylvania, the coat of arms of William Penn on the north.

The 1-mile markers in between were less elaborate, with an *M* on one side facing Maryland and a *P* on the other facing Pennsylvania. At the time of the survey, Delaware was part of Pennsylvania, so stones on the border between Maryland and Delaware also had the *M* and *P* designations. (All markers were rectangular and four sided, with blank sides between the two that were engraved.)

A search for Mason-Dixon Line markers can sometimes seem like a wild goose chase, with directions to walk across fields and wooded lots.

Locals near the line will give anecdotal descriptions of where they think there is a marker or where they heard there is a marker. But many people have never actually seen one. Markers are often on private property, deep in the woods, or in a briar patch; and they are frequently damaged, worn, or moved. Yet oftentimes the difficulty of the search makes the prize that much more precious when you finally find one.

Luckily, within driving distance of Boonsboro are two easily accessible "crown" markers—the ones that were placed approximately 5 miles apart (with *M* and *P* markers placed in between at 1-mile intervals). Here are two original markers where they're supposed to be:

To access the first marker, head north on MD 65 and take I-70 west one exit past I-81. Take exit 24 and merge onto MD 63 north (Williamsport Pike) for 1.6 miles.

Just before the Pennsylvania border, you'll see a Maryland Historical Trust sign on the left side of MD 63 stating that the 105th marker is 500 feet to the right on private property. Look to your right at the field and beyond that a house. At the back of the house's driveway is the crown marker, with the coat of arms of Lord Baltimore on one side and that of William Penn on the opposite side. Look at it from a distance, since it is on private property and getting closer would be trespassing. A taller, more modern Maryland–Pennsylvania Mason-Dixon marker with the full name of each state on either side is located next to the road on the right side of MD 63.

After carefully pulling over to savor the moment and take photos if you are so inclined, turn right onto Mason-Dixon Road (PA 163) and then right onto US 11 south past the Hagerstown Regional Airport. Then turn left onto Longmeadow Road, left onto Marsh Pike, and right onto Marsh Road. Once on Marsh Road, look for another Maryland Historical Trust sign on the right-hand side of the road. The sign points to the 100th marker on the left-hand side of the road next to some mailboxes.

Revel in the fact that you've now seen two of these obscure markers. Next you may want to visit **Lehmans Mill.** To get there, turn around on Marsh Road and turn left onto Marsh Pike. Turn left again onto Lehmans Mill Road and cross over the one-lane bridge. Parking is on the left and the store is on the right.

Located on Marsh Run, which is a tributary of Antietam Creek about a half mile from the Pennsylvania line, Lehmans Mill was built in 1760. Power for the mill came from a dam that may have been dug by slave labor. Jacob Lehman bought the mill in 1854. It stopped operating in 1996 and the property is now used as a specialty store selling jewelry, furniture, footwear, lamps, candles, and rugs. The store is certainly in a picturesque spot.

After shopping, head back by turning around on Lehmans Mill Road and making a left onto Marsh Pike. Drive 3 miles and make a left onto MD 60 to US 40 east. You can take that toward Frederick and then Baltimore or get onto I-70 east if you prefer to follow the highway.

IN THE AREA

Accommodations

Inn BoonsBoro, 1 North Main Street, Boonsboro, 21713. Call 301-432-1188. Owner and best-selling novelist Nora Roberts has obviously dotted all her i's and crossed all her t's, for every detail is well thought out. A stay here is an indulgent treat. Web site: www.innboonsboro.com.

Attractions and Recreation

Antietam National Battlefield, Sharpsburg Pike (MD 65) or 5831 Dunker Church Road, Sharpsburg, 21782. Call 301-432-5124. Battlefield is 3,250 acres, with a view of South Mountain. Open daily year-round. Run by National Park Service. Web site: www.nps.gov/anti/.

Antietam National Cemetery, MD 65, Sharpsburg, 21782. Burials took place where the soldiers fell on the battlefield. Later they were reinterred—the Union soldiers at Antietam National Cemetery, and the Confederate soldiers in Hagerstown's Rose Hill Cemetery, Mt. Olivet Cemetery in Frederick, and Elmwood Cemetery in Shepherdstown (then in Virginia, now West Virginia).

Gathland State Park and War Correspondents Memorial Arch, 900 Arnoldstown Road, Burkittsville, 21718. Call 301-791-4767. Web site: www.dnr.md.gov/publiclands/western/gathland.asp.

Gifts Inn BoonsBoro, 16 North Main Street, Boonsboro, 21713. Call 301-432-0090. Showcases community talent, from locally made quilts and scarves to pottery and jewelry. Open daily. Web site: www.giftsinnboons boro.com.

View of Boonsboro from the second-floor porch of the Inn BoonsBoro

Lehmans Mill, 19935 Lehmans Mill Road, Hagerstown, 21742. Call 301-739-9119. Specialty shop. Closed Sun. and Mon. Web site: www.lehmans mill.com.

Monocacy National Battlefield, 4801 Urbana Pike (MD 355), Frederick, 21704. Call 301-662-3515. Site of an important Civil War battle on July 9, 1864. Open daily. Web site: www.nps.gov/mono.

Rose Hill Cemetery, 600 South Potomac Street, Hagerstown, 21740. Call 301-739-3630. Burial site for more than two thousand Confederate soldiers who were killed at nearby Antietam and South Mountain.

South Mountain State Battlefield, 6620 Zittlestown Road, Middletown, 21769. Call 301-791-4767. Web site: www.dnr.md.gov/publiclands /western/southmountainbattlefield.asp.

Turn the Page Bookstore Café, 18 North Main Street, Boonsboro, 21713. Call 301-432-4588. Charming bookstore, which carries an interesting array of books and gifts, sells coffee, and has a room that carries everything Nora Roberts currently has in print, including books written under the pseudonym J.D. Robb. Nora Roberts and other authors have scheduled book signings. Open daily. Web site: www.ttpbooks.com.

Dining/Drinks

Icing Bakery, 7 North Main Street, Boonsboro, 21713. Call 301-432-5068.

Nutter's Ice Cream, 100 East Main Street, Sharpsburg, 21782. Call 301-432-5809. Old-fashioned ice cream parlor.

Old South Mountain Inn, 6132 Old National Pike, Boonsboro, 21713. Call 301-432-6155. Next to the Appalachian Trail. Dinner and Sun. brunch; closed Mon. Reservations a good idea. Web site: www.oldsouth mountaininn.com.

Palettie, 1 South Main Street, Boonsboro, 21713. Call 301-432-0500. Across the street from Inn at BoonsBoro. Open for dinner except Mon. and Tues. Web site: www.palettie.com.

Potomac Street Creamery, 9 Potomac Street, Boonsboro, 21713. Call 301-432-5242. Serves Hershey's ice cream; most popular flavors are reportedly butter pecan and mint chocolate chip.

Red Byrd Restaurant, 19409 Shepherdstown Pike, Keedysville, 21756. Call 301-432-6872. A bit of a dive and local favorite with a cheerful atmosphere and decent country cooking. Web site: http://redbyrd.net.

Other Contacts

Hagerstown-Washington County Convention & Visitors Bureau, 16 Public Square, Hagerstown, 21740. Call 301-791-3246 or 888-257-2600. Web site: www.marylandmemories.com.

Tourism Council of Frederick County, 151 East South Street, Frederick, 21701. Call 301-600-4047 or 800-999-3613. Web site: www.frederick tourism.org.

Model train exhibit at the Hagerstown Roundhouse Museum

CHAPTER

11

Brunswick, Point of Rocks, and Hagerstown

Discovering Where the Railroads Led

Estimated length: 100 miles
Estimated time: Day trip

Getting there: From the Capital Beltway around Washington, DC, travel north on I-270. Once in the Frederick area, connect to I-70 west and take US 340/15 south. When the two split, stay on US 340. Just before crossing the Potomac River, take the exit for MD 17, turn left, and travel into Brunswick.

Highlights: The historic railroad towns of Brunswick, Point of Rocks, and Hagerstown. Coffee in a former church. Railroad museums. Former ice creamery turned restaurant. Civil War history.

The trains, the railroads, and the tracks fire the imagination as they once sparked the prosperity of many towns throughout the state. This day trip explores some of those towns that prospered on the way west.

Maryland was the birthplace of American railroading. The Baltimore & Ohio (B&O) Railroad put down the first mile of track in the United States and built the first railroad station in Ellicott City. Not long after that, the tracks were expanded westward by the B&O, Western Maryland, and other railroads. When the railroad put down tracks, people and other industries followed.

Brunswick was a traditional railroad town. There were no bars, no saloons, and no liquor stores. It was a B&O town and the company could not afford to have drunken railroad workers—that would have been too dangerous.

At first, Brunswick was named Berlin since many German settlers lived in the area. But there was the other Berlin on the Eastern Shore close to Ocean City. Having two towns in the state with the same name was confusing the post office. In the 1890s, the B&O, which delivered the mail, agreed that it was too confusing. Thus the Berlin in Western Maryland became Brunswick.

Now Brunswick is a commuter town, on a Maryland Area Regional Commuter (MARC) train route. The commuter rail line from Brunswick runs into Washington on weekdays using the restored 1891 railroad station in Brunswick for one of its stops.

To see the charming Brunswick railroad station, drive into town on MD 17, travel halfway around the circle, turn right onto North Maryland Avenue, and turn right again onto South Maple Avenue. Across the large and usually crowded (on weekdays) parking lot is an entrance to the **C&O Canal towpath,** providing an option for hiking and biking.

After looking at the pretty little train station and perhaps taking a walk on the towpath, head back up the hill on South Maple Avenue and make a left onto West Potomac Street to visit the **Brunswick Railroad Museum.** Remember to feed the meter; even in the smallest of towns like Brunswick, you can get a ticket.

The museum looks deceptively small on the first floor where there's a gift shop and the adjacent C&O Canal Visitor Center. But upstairs, there's a much larger space with rather interesting lifestyle exhibits from the heyday of railroading. For instance, there's a black wedding gown, which was typical for its time; women wore their best dress to get married and it was easier to keep black clean, so it could be worn more than once. Queen Victoria of England was the first woman to wear a white wedding gown. In the 1920s, it became more common for affluent women to wear white; but the idea that white exemplified virginity came later.

After your trip into railroad-era history, you may want coffee or something to eat. For an authentic local experience, drive farther up West Potomac Street and stop at **Beans in the Belfry,** where you can get coffee and a snack in the former church turned café.

Afterward, drive approximately 2 miles on MD 17 to the traffic circle, make the first right at the circle onto MD 464, and travel to MD 15 south, where using your headlights is mandatory, even during the day. Just before the bridge crossing the Potomac River into Virginia, follow the sign and turn left into **Point of Rocks**, which is little more than a small village or hamlet. But as they say in real estate, it's location, location, location. Point of Rocks is situated on the edge of Maryland next to the Potomac River. Historically, this was a strategic location.

Originally called Trammelstown, Point of Rocks was an important transportation hub in the 1800s. The B&O Railroad built tracks and a picturesque train station, shepherding people and goods through the tiny village. The C&O Canal was also built alongside the railroad here after much controversy and competition between the two modes of transport. The railroad ended up winning the "canal versus railroad" battle for transportation supremacy, only to eventually give ground to the interstate highway system in the 20th century.

During the Civil War, the B&O Railroad line was vital for transporting supplies and for communications. However, the railroad was forced to suspend rail traffic through Point of Rocks upon threat of Confederate takeover and interference. As Lee's army advanced to Gettysburg, Point of Rocks was a major crossing point between Confederate Virginia and the conflicted border state of Maryland.

Point of Rocks also flooded several times, notably during Hurricane Agnes in 1972. Now the village is a commuter rail hub like Brunswick. MARC trains, which run throughout Maryland and into Washington, DC, travel to Point of Rocks.

Besides the commuter rail station, which operates out of the historic B&O Railroad station, as does the one in Brunswick, there's an entrance to the C&O Canal towpath at Point of Rocks as well as a liquor store, gasoline station, pizzeria, and deli. It's not much to see, but it has the authentic, "real" quality of a place that has seen floods and the movement of Civil War troops.

Specifically, Confederate general "Stonewall" Jackson captured locomotives and train cars here in 1861.

Then, in June 1863, rumors of a Confederate reinvasion halted most trains except for the mail train to Harpers Ferry and a train to **Frederick**. On June 17, Confederate cavalry attacked Union troops at nearby Catoctin

Former B&O Railroad station in Point of Rocks

Station and other troops captured a military train at Point of Rocks. Four military trains with provisions for the Union troops at Harpers Ferry managed to pass safely before this train (the last of the convoy) was captured at Point of Rocks after the train escaped the earlier attack at Catoctin Station. Anxiety grew in nearby Frederick and the surrounding countryside.

In 1864, the "Calico Raid," otherwise known as the "Crinoline Raid," took place against the Union garrison here; the raid was conducted by Confederate leader John Mosby.

By now, you'll be ready for lunch, so head back on US 15 north, which becomes US 340/15 as it nears Frederick. In the Frederick area, take the exit for US 40 west and turn left to get onto Alternate US 40 toward Braddock Mountain. This will take you right into the town of **Middletown,** for its Main Street is on the Historic National Road (US 40A). Look for the **Main Cup** and once you see it, you've found a great casual place for a meal. The Main Cup was originally an ice creamery, but now it is a restaurant that draws patrons from several nearby towns. The shrimp salad sandwich is excellent, as are most of the other dishes.

After Antietam, many wounded soldiers were nursed in Middletown as well as elsewhere throughout the region. Union troops came through the town both going to and returning from the Battle of Gettysburg. And Confederate general Jubal Early demanded and received a $1,500 ransom in exchange for not burning down the town in 1864; General Early exacted larger ransoms from Frederick and Hagerstown.

After lunch, continue on US 40A toward **Hagerstown,** which was a much larger railroad town, with the B&O, Western Maryland, and Norfolk and Western railroads having operations there. In Hagerstown, everyone had someone in the family who worked for the railroads; the railroads were by far the biggest area employer, especially in their heyday from the 1880s to 1940s. Freight was always more important than passenger service; eventually trucking on interstate highways took over a great deal of that business. Passenger service was phased out too as everyone acquired automobiles, leading to the further decline of the railroads.

There are three train museums in Hagerstown. The **Hagerstown Roundhouse Museum** is by the train tracks and has wonderful model trains, as well as memorabilia. To get there, continue on Alternate US 40 (the National Pike) to Funkstown. Turn left onto US 40 west and follow it into Hagerstown for 2.9 miles. Turn left onto North Burhans Boulevard (US 11) for 0.5 mile and then turn left onto Elgin Boulevard (US 11) for 0.1 mile. The museum is on your left after you travel under the overpass, where there is a train mural.

Just down the street from the Roundhouse Museum is the **Train Room.** It is a large store that sells miniature train supplies and memorabilia. The attached museum is geared to model railroading enthusiasts, especially Lionel fans.

The third Hagerstown railroad museum is in **City Park.** To get there, continue on North Burhans Boulevard (US 11) north until you reach the intersection with Virginia Avenue. Turn left onto Virginia Avenue for 0.4 mile, then make a left turn onto City Park Drive. The **Hagerstown Railroad Museum** is straight ahead about 0.3 mile. This museum is great, for you can climb on board a steam locomotive actually used by the Western Maryland Railroad. This gives you a chance to see what it would have felt like to be a fireman or engineer aboard an actual train, with the hot fire of the coal-fed engine beside you. Cabooses are on-site too; kids usually love them.

Jonathan Hager founded Hagerstown, along with other German settlers from Pennsylvania. It was originally called Elizabeth Town and it became Hager's Town in 1813. The National Road and the railroad helped define the city as a transportation hub. Its strategic location made it important during the Civil War.

Confederate troops under the command of General Robert E. Lee occupied the city twice, the first time before the battle of Antietam in 1862. Then in July 1864, the South once again occupied Hagerstown, ransoming the city in exchange for not setting fire to it. The sum of money paid to keep the city from burning was $20,000. Many believe that the Confederate general made a mistake on the price he demanded, leaving off a zero, for nearby Frederick's ransom was 10 times as much at $200,000.

After visiting the Hagerstown train museums, drive to the **Black Eyed Susan** in **Williamsport** for dinner. Williamsport is a suburb of Hagerstown and it was once considered as a possible location for the nation's capital. But Washington was considered more advantageous, so now Williamsport is merely a suburb of Hagerstown, with this excellent restaurant.

To get to the Black Eyed Susan, follow City Park Drive back to Virginia Avenue. Turn right onto Virginia Avenue for 0.2 mile. Make a slight left to stay on Virginia Avenue and continue for 2.6 miles. The Black Eyed Susan is on the right.

After dinner, make a left back onto Virginia Avenue and continue straight for 1.2 miles. Make a right turn onto Halfway Boulevard for 0.8 mile. Turn right onto Maryland Avenue and turn right again onto the entrance ramp for I-70 east to I-270 south back to the Washington area. Your day of riding the rails vicariously is done.

IN THE AREA

Accommodations

Homewood Suites Hilton, 1650 Pullman Lane, Hagerstown, 21740. Call 301-665-3816. An all suite property; each room has a living area and kitchen as well as a bedroom and bath. Web site: www.hagerstown .homewoodsuites.com.

Attractions and Recreation

Brunswick Railroad Museum (with C&O Canal Visitor Center), 40 West Potomac Street, Brunswick, 21716. Call 301-834-7100. The C&O Canal Visitor Center is adjacent to the gift shop on the first floor. Open Fri. through Sun. Web site: www.brm.net.

Brunswick Railroad Station, 100 South Maple Avenue, Brunswick, 21716. Restored 1891 B&O Railroad station, now serving MARC commuters to Washington, DC.

Hagerstown Roundhouse Museum, 300 South Burhans Boulevard (US 11), Hagerstown, 21741. Call 301-739-4665. Exhibits and miniature trains bring out the child in visitors of all ages. Open Fri. through Sun.; closed Jan. and Feb. Admission fee. Web site: www.roundhouse.org.

Jonathan Hager House and Museum, 110 Key Street (in City Park), Hagerstown, 21740. Call 301-739-8393. Built in 1739 over a spring to protect the founding family's water supply from Indian attacks, this house reportedly has its share of ghosts; mysterious footsteps on the stairs, rocking chairs moving unassisted, and apparitions have all been reported.

Inside there's a weasel (an old-fashioned device used to measure spun yarn) like the one mentioned in the nursery rhyme that says, "Pop goes the weasel." Open Apr. through Dec.; closed Mon. Admission fee. Web site: www.hagerhouse.org.

Hagerstown Railroad Museum at City Park, 525 Highland Way, Hagerstown, 21740. Call 301-739-8393. Chance to sit inside a steam loco-motive and learn how it operates; there's nothing like it. Open May through Oct.; closed Mon. Web site: www.hagerstownmd.org.

Hagerstown Suns, Municipal Stadium, 274 East Memorial Boulevard, Hagerstown, 21740. Call 301-791-6266. Unlike other minor league teams in the state, which are affiliates of the Baltimore Orioles, this one is a farm team for the Washington Nationals. Willie Mays played his first profes-sional baseball game here in 1950, and he was the first African American to play here in a minor league game. Mays went on to play with the New York and San Francisco Giants, and the New York Mets. Apr. through Sept. Web site: www.hagerstownsuns.com.

Prime Outlets at Hagerstown, 495 Prime Outlets Boulevard (I-70 West, exit 29), Hagerstown 21740. Call 301-790-0300 or 888-883-6288. Open daily. Web site: www.primeoutlets.com/hagerstown.

South Mountain Creamery, 8305 Bolivar Road, Middletown, 21769. Call 301-371-8565. This small family dairy farm provides an opportunity to watch cows being milked or to bottle-feed the calves. You won't get a chance to actually watch ice cream being made, but you can buy a cone at the on-site store. It's not the best ice cream around, but the farm animals are fun and educational for kids. Web site: www.southmountaincreamery .com.

The Train Room, 360 South Burhans Boulevard, Hagerstown, 21740. Call 301-745-6681. Open daily except Wed. Paid admission for the muse-um but not for the store. Web site: www.the-train-room.com.

Walkersville Southern Railroad, 34 West Pennsylvania Avenue, Walk-ersville, 21793. Call 301-898-0899 or 877-363-9777. Young children often enjoy a ride on the short line train that leads from here on the old Penn-sylvania Railroad track north of Frederick. There's an engine, an open car,

For those who worship their coffee: Beans in the Belfry is in a former church.

two coaches, and a caboose on the excursion train, which travels 4 miles. Take US 15 north from Frederick; turn right (east) onto Biggs Ford Road and proceed 2 miles to the station on the right. Open May through Oct. Web site: www.wsrr.org.

Dining/Drinks

Beans in the Belfry, 122 West Potomac Street, Brunswick, 21716. Call 301-834-7178 or 877-437-2233. Coffee shop and café inside a hundred-year-old former historic restored church; the arched stained-glass windows are still there. Live music on weekends. Open daily. Web site: www .beansinthebelfry.com.

Big Dipper, 1033 Virginia Avenue, Hagerstown, 21740. Call 301-797-5422. Regular and soft-serve ice cream.

Black Eyed Susan, 17102 Virginia Avenue, Williamsport, 21795. Call 301-582-3116. Superb salads, including a seasonal one with fresh fruit on top. Shrimp Basilica is excellent—angel hair pasta with shrimp, pine nuts, and tomato sauce. Other seafood, fish, and burgers too. Web site: www .blackeyedsusan.biz.

Kerrigan's Corner Deli, 3710 Clay Street, Point of Rocks, 21777. Call 301-874-6133. Good place to buy some soft-serve ice cream. They rotate chocolate and vanilla, so there is only one available on any day; you get what you get. Also drinks, fried chicken, and snacks. Open daily.

The Main Cup, 14 West Main Street (US 40A), Middletown, 21769. Call 301-371-4433. Great food and pleasant service. Located on the National Road between Frederick and Hagerstown. Web site: www.maincup.com.

Other Contacts

Brunswick Main Street, 13 West Potomac Street, Brunswick, 21716. Call 301-834-5591. Web site: www.brunswickmainstreet.org.

Tourism Council of Frederick County, 151 South Street, Frederick, 21701. Call 301-600-4047 or 800-999-3613. Web site: www.frederick tourism.org.

Hagerstown-Washington County Convention and Visitors Bureau, 16 Public Square, Hagerstown, 21740. Call 301-791-3246 or 888-257-2600. Web site: www.marylandmemories.com.

Main Street Middletown. Call 301-371-6171. Web site: www.mainstreet middletown.org.

Railroad museum exhibit in Ellicott City

CHAPTER

12

Ellicott City and Frederick

Finding Where Old Is New,
from Antiques to Ghosts

Estimated length: 90 miles
Estimated time: A weekend

Getting there: From Reisterstown Road (MD 140) and I-695 (Baltimore Beltway), travel south on Reisterstown Road about 0.5 mile. Turn right onto Old Court Road west 1.9 miles to Rolling Road. Make a left onto Rolling Road and continue straight for 9 miles, crossing MD 26 and US 40. At Frederick Road (MD 144 east), turn left for Catonsville and right for Ellicott City.

Highlights: Antiques shopping, ghost stories, walking a labyrinth, ice cream, restaurant inside a gasoline station, railroad museum, tales of floods and stagecoaches.

The town of **Catonsville** is quaint, with a pretty big commercial district that includes restaurants, a post office, a library, and shops. There are also several places to indulge in ice cream, including a **Baskin-Robbins** and a shop called the **Candy Bar,** which has Baltimore-area snowballs (shaved ice in a cup with a choice of syrup), plus soft ice cream and candy.

Catonsville started out in 1810 as a place where wealthy Baltimoreans built second homes to escape the summer heat of the city. An electric streetcar to Baltimore was installed by 1890, and people other than just the wealthy were able to get there. Businesses naturally evolved to serve the summer population and eventually the town became busy year-round.

The **University of Maryland Baltimore County (UMBC)** is located nearby (since 1966), as is **Catonsville Community College** (since 1958), so Catonsville is, in large part, a college town. It is also known as "Music City," for various concerts are held throughout the year and there are a lot of music stores for such a small town.

Drive in the other direction on Frederick Road and you'll quickly cross the **Patapsco River** into the historic district of **Ellicott City,** which is one of the most appealing, bustling towns around. Ellicott City is a railroad town as well as an intense shopping and dining destination.

The town has both individual antiques stores plus a few antiques malls; one of the best is **Antique Depot,** opposite the train museum. Another great shop is **Joan Eve.** There are also boutique shops that carry new merchandise. Ellicott City blends the best of the old and the new.

The town was founded in 1772 when three Quaker brothers from Bucks County, Pennsylvania, selected the land to build a flour mill. John, Andrew, and Joseph Ellicott were the founders of Ellicott's Mills, which thrived as a milling town. When Ellicott's Mills became the first terminus of the B&O Railroad outside Baltimore in 1830, the town was on its way. The **B&O Railroad Station Museum** is worth visiting, as it has a working roundabout on which railroad cars can be turned around; it is also a designated National Historic Landmark: "Oldest surviving railroad station in America." (Baltimore's station would have been older; however it is no longer standing.)

The name Ellicott's Mills was changed to Ellicott City in 1867. Before that, the town was involved in the Civil War when Union troops retreated from the Battle of Monocacy. Homes and churches were used to house wounded Union soldiers. Given the location next to the river, many floods also affected the town over the years. In one instance, Hurricane Agnes extensively flooded the historic district in the summer of 1972, after which much of the town needed to be rebuilt.

There are numerous ghost stories emanating from this town with its storied past. Many people surmise that the residual ghosts are a result of the colorful history of the town, which includes Civil War soldiers passing through. Whatever the cause of the ghostly inhabitants, Ellicott City has some real live characters on the streets and in the shops, as well as ghostly apparitions in many of the basements, staircases, and backyards.

Some say that it is such a wonderful town, no one wants to leave—living or dead. And with so many antiques in the homes of residents and for sale in the stores, perhaps ghosts are attached to particular items from which they do not want to be parted.

One Ellicott City ghost is heard opening and shutting a door, followed by gunfire, and then someone is heard tumbling down the stairs. The origin of that story is a Confederate prisoner on a train to Baltimore who escaped near what was then Ellicott's Mills. Looking for a place to hide, the Confederate soldier opened the door to a hotel room; inside were three or four Union soldiers. The Union soldiers killed the escaped Confederate prisoner and his ghost is apparently still haunting the site, replaying the fateful event.

There's a cat ghost too, and other ghosts of dead Civil War soldiers. One dead soldier is named Al and his spirit becomes most active when young, attractive women are in the vicinity.

In Ellicott City, it pays to know where the bodies are buried. Everybody seems to have a ghost, for there is a dense concentration of homes and shops in the several blocks that make up the historic main street.

There are also two cemeteries at the top of the hill and off to the left.

FINDING AND BUYING ANTIQUES

Maryland is one of the original 13 colonies, and it is a state settled by immigrants with different backgrounds and cultures. As such, it lends itself to the discovery of old treasures that can bring new thrills. Discovering these treasures is as much fun as obtaining them, and the treasure hunt is part of the travel experience.

The question of what is an antique certainly will arise as you hunt for treasures in places around the state. One definition is that an antique is anything that is at least 100 years old; a collectible is at least 50 years old. That said, putting aside any historical significance, your interest in a particular item should help dictate whether or not it is an acquisition worthy of your money and space in your home. As the saying goes, one person's trash is another person's treasure. In reality, anything you want to collect, or that interests you, could be considered a collectible.

One of the fun things about discovering a treasure tucked away amidst other "treasures" in a crowded shop is the search for the item's history and a discussion with the shop owner over its origins, if they are known.

Farmland in the countryside between Ellicott City and Frederick

One of them is where the Ellicott family plot is located; the other is where less affluent past residents are buried.

Between Ellicott City and **Frederick,** the most scenic route is MD 144 west. If you stay on Frederick Road (MD 144) from Ellicott City and drive west, there's a gasoline service station on the way with a restaurant inside. The restaurant is **Town Grill** and the town is **Lisbon.** Town Grill is not fancy—after all, it is in the middle of a service station. But the food is fantastic and the service friendly.

Town Grill has outside smokers (for meats and fish) and they serve fantastic barbecued ribs and other smoked foods. They serve breakfast too, and with the unique setting, plus the excellent home-cooked food, this is a real treasure of a stop. Town Grill closes early, however, so stop for a late lunch or an early dinner. You can, of course, get gasoline for your car at the same time.

If you're taking the backroads between Ellicott City and Frederick, **New Market** is right on MD 144 (Old National Pike) going west from Lisbon. Alternately, take I-70, exit 62 to MD 144 west. The sign in New Market reads ANTIQUES CAPITAL OF MARYLAND, which is an exaggeration. But the town does offer opportunities to buy antiques.

New Market was founded in 1793 as a stop along the Old National Pike, of which Frederick Road is a part. When you get to New Market, this historic road becomes **Main Street.** There are numerous antiques shops on Main Street as well as on alleys intersecting Main. Parking is free on both sides of the street; just be careful of traffic. While the town of New Market is known for its antiques shops, that is less true than in previous years, in part because the shops that are still there generally open only on weekends. Still, it's worth driving through even if you're there on a weekday. At least you'll get to see this quaint little town and know what's there if you decide to come back another time.

In Ellicott City, there are more than two dozen restaurants crammed into the narrow streets of the little town. In Frederick, there are more than 30 restaurants in the downtown area alone, so there's little likelihood that you'll go hungry anywhere on this trip.

From New Market, continue on MD 144, which becomes East Patrick Street and leads directly into the heart of the historic district of Frederick. Parking in downtown Frederick is available in five garages and at meters on the street.

In Frederick, there are ghosts too, like the ones who are said to roam the **National Museum of Civil War Medicine.** One is supposedly an undertaker's assistant and another is rumored to be a railroad worker who transported dead soldiers long ago. Some ghosts are apparently emotionally attached to items on display in the museum, items that once belonged to them when they were alive.

It is not unusual for visitors to see apparitions in Frederick, especially in the alleyways and historic buildings, even if those visitors have not been visiting the bars.

Downtown Frederick has lots of antiques shops, especially in the **Market Street-Patrick Street corridor** at the heart of the downtown area. The city of Frederick is itself a relic, for it dates back to 1745 and many of its buildings have been preserved. Diverse architectural styles range from Federal to art deco. For this trip, we're mostly pursuing antiques and ghosts, but there is much else to do in Frederick, including dining and visiting historic sites.

Besides the shops in a several-block radius around Market and Patrick Streets, you can go shopping at **Everedy Square/Shab Row**—a small area

Visitors can walk a circular path in the labyrinth.

of once-dilapidated buildings that has been restored into a pleasant shopping destination. The Everedy bottle capping company used to have its business there, as did tinsmiths, wheelwrights, and others who worked in the shops on Shab Row serving stagecoaches traveling west through Frederick.

During the Civil War, Confederate general Jubal Early occupied the city; he demanded a $200,000 ransom or threatened that the invading army would set fire to the city. The city of Frederick paid the ransom in July 1864, and thus avoided being torched. Luckily the city survived, for it is filled with marvelous historic buildings, shops, and restaurants. Of particular interest is the burial site of Francis Scott Key at the entrance to **Mt. Olivet Cemetery** near **Harry Grove Baseball Stadium**.

Maryland was in truth a significant border state and it had a sizable pro-Southern population interspersed with those loyal to the Union. In fact, in 1861, Maryland was going to vote for possible secession, along with the Southern states.

To make sure the state did not vote to join the Confederacy, President Lincoln asked then Maryland governor Thomas Hicks to hold the state's

General Assembly in Frederick rather than in the more Southern-leaning Annapolis. The state legislature met in Frederick's Kemp Hall (at the corner of North Market and East Church Streets), where federal soldiers arrested several pro-Southern legislators. The arrests ensured that Maryland would not choose disloyalty to the Union; in denying the legislature a quorum, the federal government made sure that no vote could take place.

To return to the Baltimore area from downtown Frederick, take East Street to the exits for I-70, turning left onto the ramp for I-70 east. Before you reach I-695 (Baltimore Beltway), take a detour off I-70 by getting off on exit 83. Take the north ramp and merge onto Marriottsville Road. Drive 1.5 miles to the **Bon Secours Spiritual Center** on the right. While this is a religious retreat, everyone is welcome to walk the large labyrinth on the grounds.

Walking a labyrinth is said to clear the mind and provide spiritual insight. After exploring history, antiques, and ghosts, this relaxing activity is perfect. It is said to bring wisdom and peace to those who concentrate on walking rather than on other things. By walking the circular labyrinth, new paths for one's life may suddenly become apparent. In any case, it is a beautiful setting and walking the labyrinth is relaxing.

Afterward, you can easily get back on I-70 east once more toward Baltimore.

IN THE AREA

Accommodations

Inn at Buckeystown Bed & Breakfast, 3521 Buckeystown Pike (MD 85), Buckeystown, 21717. Call 301-874-5755. Victorian-style B&B. The house is large, dating back to 1899. Besides renting rooms to overnight guests, the B&B holds special "murder mystery" evenings, which seems appropriate in such a dramatic setting. Located near the Monocacy River, about 5 miles south of Frederick. During the Civil War, Confederate generals marched through the village. Web site: www.innatbuckeystown.com.

The Hotel at Turf Valley, 2700 Turf Valley Road, Ellicott City, 21042. Call 410-465-1500 or 888-833-8873. A luxurious resort with a superb restaurant and challenging golf course, the hotel also has a spa for massages and other indulgences. Web site: www.turfvalley.com.

Wayside Inn Bed & Breakfast, 4344 Columbia Road, Ellicott City, 21042. Call 410-461-4636. Rumor has it that the resident female ghost escaped when a wall was knocked out during a recent renovation. Antiques and private baths. Web site: www.waysideinnmd.com.

Attractions and Recreation

Bon Secours Spiritual Center, 1525 Marriottsville Road, Marriottsville, 21104. Call 410-442-1320. A place where people of any faith can walk the trails and the especially appealing labyrinth. Web site: www.bonsecours spiritualcenter.org.

B&O Railroad Museum: Ellicott City Station, 2711 Maryland Avenue (corner of Main Street), Ellicott City, 21043. Call 410-461-1945. Construction on this railroad station was completed in 1830, making this the oldest surviving railroad station in the United States. The first commercial railroad track ever constructed was completed by the B&O Railroad; it was 13 miles of track and it started in Baltimore and ended here in Ellicott City. The turntable was added in 1863. Now the station is a wonderful museum. Open Wed. through Sun. Web site: www.ecborail.org.

The Candy Box, 1610 Frederick Road, Catonsville, 21228. Call 410-747-5291. Ice cream, snowballs, candy, and milk shakes.

Clark's Elioak Farm, 10500 Clarksville Pike (MD 108), Ellicott City, 21042. Call 410-730-4049. Enchanted Forest fantasy figures from nursery rhymes and fairy tales that have been moved to this farm to preserve them and bring joy to new children. There's also a petting zoo, pony and hay rides, and a pine-tree maze, plus a farm stand with fresh produce out by the road. A perfect family-oriented farm, ideal for very young children. Open Apr. through Nov.; closed Mon. Web site: www.clarklandfarm.com.

Everedy Square and Shab Row, corner of East and Church Streets, Frederick, 21701. Call 301-662-4140. Three blocks in downtown Frederick, combining the historic buildings of Shab Row, which formerly serviced stagecoaches traveling west on the National Pike (now US 40), and Everedy Square, once the home of a bottle cap and kitchenware company. These buildings were condemned slums that were restored instead of being destroyed. Shops sell antiques, gifts, and other items. Web site: www.everedysquare.com.

Ellicott City's B&O Railroad Station Museum

Frederick Keys Baseball-Harry Grove Stadium, 21 Stadium Drive, Frederick, 21703. Call 301-662-0013 or 877-846-5397. Home of the Frederick Keys, minor league baseball, Class A Carolina league affiliate of the Baltimore Orioles. Web site: www.frederickkeys.com.

Historic Savage Mill, 8600 Foundry Street, Savage, 20763. Call 410-880-0918 or 800-788-6455. Close to Ellicott City, this 19th-century cotton mill has been turned into a shopping mall with antiques stores as well as other shops. Web site: www.savagemill.com.

Joan Eve, 8018 Main Street, Ellicott City, 21043. Call 410-750-1210 or 866-750-1210. Web site: www.joaneve.net.

Mt. Olivet Cemetery, 515 South Market Street, Frederick, 21701. Call 301-662-1164 or 888-662-1164. Final resting place for Francis Scott Key, author of the lyrics to our national anthem, "The Star-Spangled Banner," who was buried two times before reaching his current spot at the front of the cemetery—once in Baltimore and the second time in his family's plot elsewhere at Mt. Olivet. Thomas Johnson, the first governor of Maryland, is also buried at Mt. Olivet. Web site: www.mountolivetcemeteryinc.com.

National Museum of Civil War Medicine, 48 East Patrick Street, P.O. Box 470, Frederick, 21705. Call 301-695-1864 or 800-564-1864. An appropriate place for such a museum, as 29 hospitals were set up in Frederick's public buildings during the Civil War. Wounded were brought to the city from various battles, including Antietam and Gettysburg. Open daily. Web site: www.civilwarmed.org.

Old Glory Antique Marketplace, 5862 Urbana Pike (MD 355 South), Frederick, 21704. Call 301-662-9173. More than one hundred dealers. Open daily. Take I-70 west to exit 54 and MD 355 south.

Dining/Drinks

Acacia Fusion Bistro, 129 North Market Street, Frederick, 21701. Call 301-694-3015. Exceptional upscale food, friendly service. Trout topped with crabmeat, mashed sweet potatoes, and other excellent dishes. Web site: www.acacia129.com.

Catonsville Gourmet Market & Fine Foods, 829 Frederick Road, Catonsville, 21228. Call 410-788-0005. Excellent seafood. BYOB. Open daily for lunch and dinner. Web site: www.catonsvillegourmet.com.

Ellicott Mills Brewing Company, 8308 Main Street, Ellicott City, 21043. Call 410-313-8141. Superb beer-battered fish and chips. Web site: www.ellicottmillsbrewing.com.

Matthew's 1600, 1600 Frederick Road, Catonsville, 21228. Call 410-788-2500. French turn-of-the-century posters, comfortable booths, cheerful bar, and good food. Web site: www.matthews1600.net.

Rams Head Tavern, 8600 Foundry Street, Savage, 20783. Call 410-604-3454. Web site: www.ramsheadtavern.com.

Town Grill, 15943 Frederick Road, Lisbon, 21765. Call 410-489-5016. Smoked barbecued ribs and other meats. Located inside the Lisbon Citgo Auto Center. A unique experience. Open for breakfast and lunch, Mon. through Sat. Web site: www.lisbontowngrill.com.

Volt, 228 North Market Street, Frederick, 21701. Call 301-371-0400. Chef Bryan Voltaggio was runner-up in the 2009 season of the hit cooking show *Top Chef.* Hard to get a reservation, but the food is excellent. Web site: www.voltrestaurant.com.

Other Contacts

Baltimore County Conference & Tourism, 44 West Chesapeake Avenue, Towson, 21204. Call 410-296-4886 or 800-570-2836. Web site: www.enjoybaltimorecounty.com.

Greater Catonsville Chamber of Commerce, 924 Frederick Road, Catonsville, 21228. Call 410-719-9609. Web site: www.catonsville.org.

Howard County Tourism and Promotion, 8267 Main Street (side entrance), Ellicott City, 21043. Call 410-313-1900 or 800-288-8747. Web site: www.visithowardcounty.com.

New Market Dealers Association, P.O. Box 295, Frederick, 21774. Web site: www.newmarkettoday.com.

Tourism Council of Frederick County, 151 East South Street, Frederick, 21701. Call 301-600-4047 or 800-999-3613. Web site: www.fredericktourism.org.

The Utica Mills Covered Bridge across Fishing Creek

CHAPTER

13

Thurmont and Emmitsburg

Exploring Close to Catoctin Mountain

Estimated length: 65 miles
Estimated time: Day trip

Getting there: From Frederick take Catoctin Mountain Highway (US 15) north to the junction with Old Frederick Road. Turn right onto Old Frederick Road and drive 1.5 miles to Utica Road. Make a left and follow the road to the Utica Mills Covered Bridge.

Highlights: Tour of three wooden covered bridges. Shopping for antiques. Gorgeous mountain countryside and farmland. Near Camp David in Catoctin Mountain National Park. Cunningham Falls State Park.

This is a beautiful drive, especially in the fall, for Catoctin Mountain is the easternmost part of the Blue Ridge Mountains (which are the start of the Appalachian Mountains).

The original roads in this part of the state were Indian trails used by the Iroquois and Algonkians. As European traders and pioneers made forays into the area, the roads became more defined. In the mid-1700s, there was a turnpike from **Frederick** to **Emmitsburg.** Later, the roads paralleling the mountain were used by Civil War troops on their way to Gettysburg for the battle; afterward, the troops brought dead and wounded from both sides to

Emmitsburg, **Thurmont,** and Frederick. Many churches and other buildings were commandeered as makeshift hospitals.

The first stop of the day is the **Utica Mills Covered Bridge,** which dates to 1850. It originally spanned the nearby **Monocacy River.** However, it was washed away in 1889 during a severe storm. The current location of this 101-foot bridge spans **Fishing Creek;** this reconstructed version dates to 1891.

After seeing the bridge, return to Old Frederick Road and make a left onto it. Drive 3.9 miles to a stop sign. Make a left onto MD 550 and drive 0.4 mile before taking the right fork in the road back onto Old Frederick Road. Travel 2 miles and make a left into the parking lot. You're now at **Loy's Station Park,** where there's another barn-red, one-lane wooden covered bridge spanning **Owens Creek.** This bridge was originally constructed in 1880. It was then rebuilt after being almost destroyed by fire in 1991.

Between storms, fires, and normal wear and tear, one starts to understand why so few wooden covered bridges are left. In Maryland, there are only about six still standing, three of which you'll see on this drive. In the mid- to late 1800s, covered wooden bridges were commonplace. Time, disasters, and neglect have caused most of them to disappear.

After you've spent some time admiring the 90-foot-long **Loy's Station Covered Bridge,** and perhaps walked around the park, make a left out of the parking lot and drive through the bridge. Proceed 0.3 mile to a stop sign and make a left onto Rocky Ridge Road (MD 77). Travel 2.7 miles and make a right onto Apples Church Road in Thurmont. Go straight for 1.6 miles and drive through the bridge to a parking lot on the left. Built in 1856, the **Roddy Road Covered Bridge** over Owens Creek is the smallest of the three, measuring only 40 feet long.

After you look at this third bridge, make a right onto Roddy Creek Road from the parking lot. Drive 0.4 mile to a stop sign at the junction of US 15. Take US 15 north about 4 miles to Emmitsburg, which is right before the Pennsylvania border. Emmitsburg is only a few miles south of Gettysburg, Pennsylvania, and the important battlefield there. During the Civil War, troops passed through the town both coming and going.

While the Battle of Gettysburg took place on Pennsylvania soil, the invasion of the Confederate troops and their retreat, as well as the Union Army advance, took place through Maryland, so many places were affected by that pivotal battle. Emmitsburg, for instance, was a Union supply depot. And while the battle at Gettysburg lasted three days, the campaign took 35 days, with most of the advance and retreat occurring in Maryland. The intent by Confederate general Robert E. Lee was to carry the fighting across the Mason-Dixon Line into Pennsylvania. This part of Maryland was greatly impacted.

The Union, on the other hand, was determined to stop the "invaders" while also protecting Baltimore and Washington, DC. With both armies on Maryland soil so much, and with federal suspension of some civil rights in areas where martial law was imposed, plus citizens arrested for disloyalty to the government, this area was directly affected. Additionally, there were destroyed roads, crops, and fences and stolen livestock and crops. Marylanders, while seeing less suffering and deprivation than the Southern states, still had its share of challenges during the Civil War.

Emmitsburg is also the location of the **National Shrine of St. Elizabeth Ann Seton,** the first American-born canonized saint, plus an antiques mall, a few restaurants, and **St. Joseph's and Mount St. Mary's Colleges.** Mount St. Mary's is the oldest independent Catholic college in the United States. Entrance to the Elizabeth Seton shrine is free, and you can wander around the campus or just drive through.

By now you are probably hungry and ready for lunch. The best place in these parts to have lunch is the **Carriage House Inn;** its country cooking has kept it around for many decades.

You may want to note that **Camp David,** the presidential retreat, is not far away; in fact, you'll drive close to it on the ride back to Frederick. One night, President Bill Clinton and First Lady Hillary Clinton dined at the Carriage House Inn with friends; the president and first lady were a hit with the employees, for they took time to speak with everyone and take photos.

The Carriage House building dates to 1857 and it was formerly a feed store, a broom factory, and a bus depot; it became a restaurant named the White House in 1943 (appropriate, as it turns out, for a location near the presidential retreat). When the current owners bought the building in 1986, they renamed it the Carriage House. The owner is proud that his restaurant goes through 150 pounds of crabmeat weekly, much of which goes into crabcakes; the restaurant is also known for its steaks.

After lunch, leave your car in the lot and take a walk up South Seton Avenue to East Main Street and turn right. You'll find a quaint small-town scene plus a store called **Antiques Folly** that is open on weekends. Behind the store, about a block away on Chesapeake Avenue, is **Emmitsburg Antique Mall,** a much larger place where you can see antiques from a variety of dealers; the antiques mall is open daily.

After you've looked around at some antiques, walk back to where your car is parked at the Carriage House Inn. From there, turn right onto South Seton Avenue, travel straight, and exit onto US 15 south. You'll soon come to two scenic lookouts and you can pull over at one or both to look at the scenery. You'll also pass **Catoctin Mountain Orchard,** where you can stop to buy fresh fruit and pies. The orchard's entrance and parking lot are on the right-hand side of the road; access it directly on and off US 15.

You'll soon pass by the small town of Thurmont, which was called Mechanicstown at one time. The Western Maryland Railroad made the town important both for shipping and for Baltimore residents looking to find a mountain escape from hot summers in the city. The railroad found that the name Mechanicstown was too long for signage and fare cards, so the name needed to change. Some residents wanted to call the town Blue Mountain City instead, but that name was also too long. Thurmont, perhaps short for "through the mountains," was eventually chosen.

Thurmont is perhaps best known for its proximity to Camp David, the U.S. presidential retreat, which is close enough to Washington that it is used often. Foreign heads of state are sometimes asked to join the president there.

In 1942, soon after the beginning of direct American involvement in World War II, President Franklin Roosevelt adopted a portion of Catoctin Mountain as his personal mountain retreat, naming it "Shangri-La." In the 1950s, President Dwight Eisenhower renamed the retreat Camp David after the name of both his father and grandson, David Eisenhower. Numerous foreign leaders have visited various presidents at the retreat, including Winston Churchill in 1943, Soviet leader Leonid Brezhnev during Nixon's time in office, Anwar el-Sadat and Menachem Begin at the Camp David Accords during Jimmy Carter's administration in 1978, Margaret Thatcher, who visited Ronald Reagan, and Ehud Barak and Yasser Arafat at the 2000 Middle East Peace Summit held by Bill Clinton.

You cannot visit Camp David but you can drive close to it by taking the MD 77 west exit off US 15. That will put you on the road between two impressive parks, for MD 77 is the dividing line between a federal park, **Catoctin Mountain National Park,** on one side and a state park—**Cunningham Falls State Park**—on the other.

To enter the national park, drive about 2.5 miles on MD 77 and turn right onto Park Central Road. When you're driving through the park on this road, you'll drive by Camp David, but it's not marked and there are no obvious guards. However, photos are not allowed and if you try to take them, you will be swarmed by security officers, so it's not advisable. There

The Camp David Museum is open to the public inside the Cozy Restaurant.

is a sign that says CAMP #3 at the seemingly unimposing entrance. But don't be fooled. Camp David is actually a naval facility guarded by marines and restricted from traffic by car, foot, and air.

Also in the national park is **Blue Blazes Whiskey Still**—a remnant of the Prohibition era. To see it, park in the first lot by the visitors center as you turn right onto Park Central Road. There's a marked trail for the whiskey still. Just follow that on foot about a half mile to the site where the illegal brew was made all those years ago. There are other trails for hiking and biking in the park as well.

Across the road, Cunningham Falls State Park has a waterfall, a lake, and miles of hiking trails. The state park is also where the **Catoctin Furnace** is located; federal troops cast cannonballs at the furnace during the War of 1812.

After visiting the parks, head back on MD 77 east. Make a right on Tippin Drive and turn right onto Frederick Road. You'll want to stop at a restaurant and inn called **Cozy.** Inside is a museum about Camp David. It's free, it's wonderful, and it's worth stopping there to learn a bit of the history behind the important presidential retreat. This also makes a good stop for dessert and iced tea.

To complete your round-trip excursion, make a right turn from the Cozy parking lot onto Frederick Road, go to the traffic signal, turn right, travel under US 15, and make the first left onto the exit ramp for US 15 south.

IN THE AREA

Accommodations

Sleep Inn & Suites of Emmitsburg, 501 Silo Hill Parkway, Emmitsburg, 21727. Call 301-447-0044. Web site: www.sleepinnemmitsburg.com.

Attractions and Recreation

Antiques Folly, 20 East Main Street, P.O. Box 1017, Emmitsburg, 21727. Call 301-447-5967. Antiques shop. Open Sat. and Sun.

Camp David Museum at Cozy, 103 Frederick Road (MD 806), Thurmont, 21788. Call 301-271-7373. This museum is located inside the Cozy Restaurant, which has served presidents, foreign leaders, and the press. This is a museum with free admission, and it's well worth stopping to see the memorabilia about nearby Camp David. You need not eat at the restaurant or stay at the inn to visit the museum. Web site: www.cozy village.com.

Catoctin Mountain National Park, 6602 Foxville Road, Thurmont, 21788. Call 301-663-9388. Camp David is here, hidden away and well guarded. Web site: www.nps.gov/cato.

Catoctin Mountain Orchard, 15036 North Franklinville Road, Thurmont, 21788. Call 301-271-2737. You can pick your own fruit (seasonal).

There are also fruit pies and canned peaches. Closed Feb. through Apr. Web site: www.catoctinmtorchard.com.

Cunningham Falls State Park, 14039 Catoctin Hollow Road, Thurmont, 21788. Call 301-271-7574. Web site: www.dnr.state.md.us/public lands/western/cunninghamfalls.html.

Emmitsburg Antique Mall, 1 Chesapeake Avenue, Emmitsburg, 21727. Call 301-447-6471. More than 120 booths with antique furniture, linens, quilts, glassware, china, toys, tools, and collectibles. Open daily. Web site: www.emmitsburg.net/antique_mall/.

Loy's Station Covered Bridge, 3800 Old Frederick Road, Thurmont, 21788.

National Shrine of St. Elizabeth Ann Seton, 333 South Seton Drive, Emmitsburg, 21727. Call 301-447-6606. Site honors first American-born canonized saint; she is buried in the basilica on the shrine's campus. Closed Mon. Free admission and parking. Web site: www.setonshrine.org.

Roddy Road Covered Bridge, 14760 Roddy Road, Thurmont, 21788.

Utica Mills Covered Bridge, Old Frederick Road (over Fishing Creek), Utica, 21701.

Dining/Drinks

Carriage House Inn, 200 South Seton Avenue, Emmitsburg, 21727. Call 301-447-2366. Country cooking and friendly service. Lunch and dinner daily plus Sunday brunch. Web site: www.carriagehouseinn.info.

Red's Tavern, 135 Chesapeake Avenue, Emmitsburg, 21727. Call 301-447-6749. Baseball memorabilia, a pool table, and comfort food at reasonable prices.

Other Contacts

Tourism Council of Frederick County, 151 South East Street, Frederick, 21701. Call 301-600-4047 or 800-999-3613. Web site: www.frederick tourism.org.

Union Mills Homestead, which belonged to the Shriver family

CHAPTER

14

Westminster, Taneytown, and Sykesville

Exploring Historic Small Towns of Central Maryland

Estimated length: 80 miles
Estimated time: Day trip or overnight

Getting there: From Baltimore, take Reisterstown Road (MD 140) north into the town of Reisterstown. Continue on MD 140 by bearing left at the fork toward Westminster rather than toward Hampstead on MD 30. From Westminster, continue on MD 140 to Taneytown. To reach Antrim 1844 (Country House Hotel), bear right at the traffic circle and drive 1 mile, turning left onto Trevanion Road and turning right after about 150 feet at the first set of brick pillars.

Highlights: Wonderful small-town life. Down-to-earth restaurants. Charming main streets. Rural farm countryside. Antrim 1844 Inn. Taneytown. Westminster. Union Mills. Sykesville. Mason-Dixon crown stone right over the Pennsylvania border.

Whether you decide to stay overnight or just make this a day trip, it can be wonderfully relaxing and restful to drive through some of the small towns of Central Maryland, pursuing a bit of history, a bit of shopping, a few enticing meals, and a pretty drive.

We'll start out in Taneytown since that's where **Antrim 1844** is located. If you're staying overnight, there's no place as dramatic as Antrim.

Taneytown (locals pronounce it "Tawney-town") is only about 15 minutes south of Gettysburg across the Mason-Dixon Line into Pennsylvania, so the Civil War figures prominently in its history. Not surprisingly, the town was in the midst of concentrated troop activity in 1863. It is rumored that Union generals watched some of the troop movements from the widow's watch at the top of the mansion at Antrim 1844.

Antrim…you drive up and immediately fall in love with the stately, antebellum-style grounds and mansion. In spite of its military history, one is surprised to find this elaborate and genteel place in rural Maryland. One has the sense that the entire B&B—a misnomer, for it's more like an inn—belongs in Charleston, South Carolina, not north-central Maryland, but here it is, catering to affluent guests and those enjoying special occasions.

The mansion on the former plantation dates back to 1844. You'll think you're at Margaret Mitchell's Tara when you first arrive—it's quite the impressive setting with a gracious, Southern kind of style.

Antrim's **Smokehouse Restaurant** features a five-course dinner, and the inn provides guests with an opportunity to play croquet, selections from an extensive wine cellar, formal gardens, and a pool.

Then at dinner, one goes from *Gone With the Wind* to the scenes in *Pretty Woman* when the Julia Roberts character learns which utensils to use—there are, after all, three forks, a butter knife, a soup spoon, three knives, and a tiny spoon. Antrim specializes in five-course meals, after all. More food than one needs, to be sure. And this formality is certainly not what one would expect from the small-town life all around. And yet, here it is.

The service at the inn seems impeccable too, if a bit odd or quirky. One discovers that the housekeepers hide the soap inside towels, even though most guests have trouble finding it. Soap left out is considered messy, guests are told. There must be a rule book or etiquette manual somewhere, but it is not shared with those guests.

If you like wine, the wine cellar is remarkable; you can ask for a tour. The lawns are manicured, the gardens impeccable, fountains working, rosebushes clipped, croquet lawn at the ready.…

But back to small-town life.

From **Westminster,** since it is centrally located in this area, take MD 140 west and go left on MD 31 to New Windsor and then right on MD 75 to the village (or town, depending upon your definition) of Union Bridge.

Formal gardens on the grounds at Antrim 1844

Here the **Western Maryland Railroad Museum** sits beside some tracks in a former train station. The collection and its story are fascinating if you like trains, if you like Maryland, and if you like history. It is run by volunteers and you can tell this is a passion for them.

Next on the itinerary is **Union Mills Homestead** on Littlestown Pike (MD 97) about 6 miles north of Westminster. This homestead formerly belonged to the Shriver family and it is where they had a mill, a tannery, and a canning business.

During the Civil War, one faction of the family had Southern sympathies, the other Northern. Ironically, it was the Northern faction that owned slaves, whereas the Southern faction had none. Troops from both armies, on their way to nearby Gettysburg and the upcoming battle there, ate a meal at the homestead and just missed each other by a day. It's an interesting historic home, with pretty grounds and a great story—appealing even for those who are not fans of historic houses, as this one is special.

The town in this group that's in a different direction is **Sykesville,** which is south on MD 32 from Westminster. Sykesville is charming, and you should include a drive there in this trip. Everything is relatively close, so it doesn't matter too much in what order you visit the different towns and surrounding communities. Deciding how long you want to spend at each place will help determine your exact itinerary. There is enough to explore for several days, or you can consolidate the trip into one day and visit fewer places.

In Sykesville, there's a former B&O railroad station, built in 1884 to replace the one that was lost in a major 1868 flood. Eventually the B&O stopped providing passenger service on the "Old Main Line," which included Sykesville, and freight

LOCAL FACTS

Francis Scott Key, author of "The Star-Spangled Banner," was born about 5 miles from Taneytown. He taught Sunday school to slave children at the church across the street from the courthouse in Westminster.

If you want to sound like a local, pronounce the name of Westminster "West-min-ster" with three syllables, not "West-min-i-ster" with four.

service was finally discontinued in 1982 too. Now the station houses a wonderful restaurant called **Baldwin Station.** The train tracks between the restaurant and the Patapsco River are still used, but there are no more train stops in Sykesville.

Before the flood, the town was on both sides of the Patapsco River. But the flood wiped out a good portion of the town and when it was rebuilt, it was rebuilt strictly on the higher north side.

During the Civil War, the Confederates burned the bridge in Sykesville, destroyed the telegraph line, and tore up the railroad tracks, thus attempting to eliminate communication between the Union generals and Washington, DC, during the crucial time leading up to the Gettysburg battle in June 1863.

Sykesville is like a smaller, less busy Ellicott City, with stone buildings and a former train station near the railroad, a hill heading up the main street, and various boutique shops, including those selling books and antiques. Sykesville is like Ellicott City was years ago, before Ellicott City became well known and more crowded.

Traveling the two-lane back roads throughout the middle of the state reveals beautiful farmland as well as an assortment of towns and villages with country stores, homemade ice cream, and unique architectural finds. This area saw Civil War skirmishes, though no major battles, and provided transportation routes for armies and their supplies, as well as points of communication from high spots on hilltops and church steeples.

The railroad played an important part in the growth and prosperity of these little towns. Transportation, supplies, and affluence were all things that came on the rails. The B&O out of Baltimore supplied the Union Army during the Battle of Gettysburg and at other times, with a depot located in Westminster and another in Sykesville. The Western Maryland Railroad also connected Westminster with points farther west and north, so it too played a role.

Manchester is a pretty little town near the Pennsylvania border. To get there from the town of Reisterstown, take MD 30 north; from Westminster, take MD 27 north.

In order to see a Mason-Dixon Line "crown stone" (5-mile marker) in this area, travel north through Manchester on MD 30 and continue into Pennsylvania, crossing the Mason-Dixon Line.

Once across the Maryland–Pennsylvania border, make an immediate

Mason-Dixon Line crown marker north of Manchester

left onto Garrett Road, go around 0.2 mile, and make the first left on what seems like an unmarked street, but is in reality a continuation of Garrett Road—the sign has just been moved. The Mason-Dixon crown stone is on the left surrounded by a wrought-iron fence.

IN THE AREA

Accommodations

Antrim 1844 Country House Hotel, 30 Trevanion Road, Taneytown, 21787. Call 410-756-6812 or 800-858-1844. Web site: www.antrim1844 .com.

Atlee House Bed & Breakfast, 120 Water Street, P.O. Box 430, New Windsor, 21776. Call 410-871-9119. The beds are comfortable, the house is clean and historic, and the proprietors are friendly and accommodating. There's not much in New Windsor itself, but from here it's an easy ride throughout the area. Web site: www.atleehousebb.com.

Best Western Westminster Hotel, 451 WMC Drive, Westminster, 21158. Call 410-857-1900. The NFL Ravens football team lives here during training season when they practice in the stadium at McDaniel College just down the street. Web site: www.bestwesternwestminster.com.

Attractions and Recreation

Historical Society of Carroll County, 210 East Main Street, Westminster, 21157. Call 410-848-6494.

McDaniel College, 2 College Hill, Westminster, 21157. Call 410-848-7000. Home of the Baltimore Ravens summer camp and NFL training ground; the college and nearby town sport the Ravens' colors—purple and black—when the team members are in town. McDaniel was formerly Western Maryland College. Before that, it was "the Hill," an important vantage point during the nearby Battle of Gettysburg just north in adjacent Pennsylvania. Web site: www.mcdaniel.edu.

Union Mills Homestead, 3311 Littlestown Pike (MD 97), Westminster, 21158. Call 410-848-2288 or 800-272-1933. Open June through Aug.; closed Mon. Web site: www.unionmills.org.

Western Maryland Railway Historical Society Museum, 41 North Main Street, Union Bridge, 21791. Call 410-775-0150 or 410-356-9199. Web site: www.moosevalley.org/wmrhs.

Dining/Drinks

Baldwin's Station, 7618 Main Street, Sykesville, 21784. Call 410-795-1041. This restaurant is exciting because of its location in a former B&O railroad station building. Sykesville is quaint and appealing, like a smaller Ellicott City. Food is good too, and there's frequently live music. Web site: www.baldwinstation.com.

Greenmount Station, 1631 North Main Street, Hampstead, 21074. Call 410-239-0063. Excellent food and friendly service. Terrific old historic photos in one of the several dining rooms. Open daily for lunch and dinner. Web site: www.greenmountstation.com.

Gypsy's Tearoom, 111 Stoner Avenue, Westminster, 21157. Call 410-857-0058. This is a charming place to have lunch or afternoon tea and the homemade scones are fantastic. There's even coffee for those who don't want to drink tea, as well as finger sandwiches and other goodies. Open daily. Web site: www.gypsytearoom.com.

Maggie's, 310 East Green Street, Westminster, 21157. Call 410-876-6868. For a landlocked location, Maggie's has superb crabcakes. These 5-ounce delights are worth a trip to Westminster, which was once a part of Baltimore County. This come-as-you-are location is lively, with a bar and several small rooms. Web site: www.maggieswestminster.com.

Other Contacts

Baltimore County Conference & Tourism, 44 West Chesapeake Avenue, Towson, 21204. Call 410-296-4886 or 800-570-2836. Web site: www.enjoybaltimorecounty.com.

Carroll County Tourism, 224 North Center Street, Room 100, Westminster, 21157. Call 410-386-2983 or 888-299-2983. Web site: www .carrollcountytourism.org.

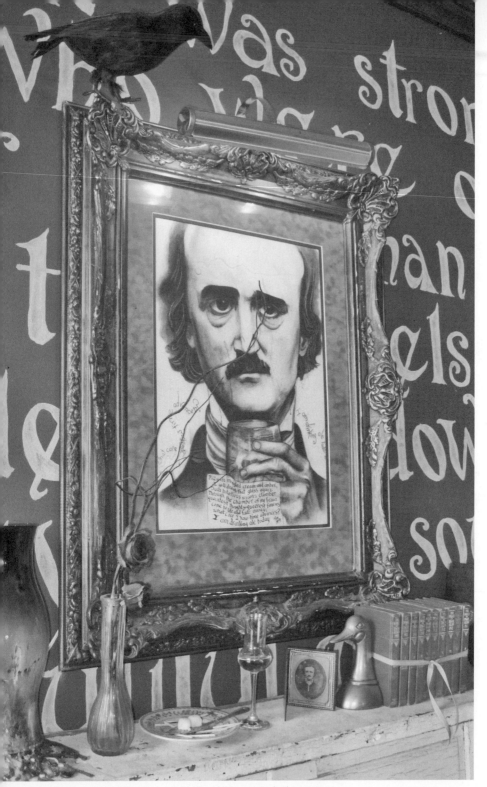

Edgar Allan Poe memorabilia at the Annabel Lee Tavern

CHAPTER

15

Hidden Baltimore— A City of Neighborhoods

Uncovering Places Even Many Locals Don't Know

Estimated length: 65 miles
Estimated time: Series of day trips or a leisurely weekend

Getting there: We'll start at Baltimore's Inner Harbor. If you're coming from the south, head north on Charles Street for a few blocks (with the Inner Harbor on your right) until you reach the intersection with Baltimore Street. There are parking lots all around and occasionally you will find on-street parking (make sure to buy a paper pass for your windshield from a nearby machine).

Highlights: B&O Railroad history. Edgar Allan Poe burial site and Annabel Lee Tavern. Former Prohibition-era speakeasy. Fells Point. Mt. Vernon. Little Italy. Johns Hopkins University.

Baltimore has tremendous character, for it is actually a conglomerate of small ethnic neighborhoods with many hidden surprises. This is a real hometown with true camaraderie, a port city that figured prominently in the immigrant history of the United States as well as the War of 1812 and the Civil War, a city that prides itself on retaining its unique, somewhat insular, lifestyle.

It is a city of eccentricity, where water is pronounced "warder" and waitresses call everyone "Hon," a traditionally blue-collar town with a new sophistication. "Bawlmer" is a place where there are art museums and a symphony, yet everyone loves to sit outside in the summertime and eat messy hardshell steamed crabs covered in Old Bay Seasoning.

Baltimore is often ranked among the "most fun" cities in America; good "eats" are no doubt a major part of that designation. Seafood lovers will be enthralled, while the city's various ethnic populations also provide dishes from their special heritages. Food here is not pretentious, but it is superb.

Everyone thinks of the **Inner Harbor** when they think of Baltimore. While this is, indeed, part of the Baltimore experience, there's much more. There's **Fort McHenry,** of course, the sight of which inspired Francis Scott Key to write the words to the "Star-Spangled Banner," our national anthem. Downtown Baltimore also has the **Bromo Seltzer clock tower.** Although not the size of London's Big Ben, the Bromo Seltzer tower is to Baltimore what Big Ben is to London, a recognizable symbol on the city's horizon.

Baltimore is part Northern and part Southern in its emotional ties; and it is home to major league sports teams and fans. If there's a game in town, it's well worth the cost of tickets to attend an Orioles or Ravens game. Even if there's no game, talk of sports abounds. Visitors are inevitably pleased with what they find in this place that is often referred to as "Charm City," especially if they search out Baltimore's "hidden" charms.

The route we'll take encompasses the **Charles Street** corridor, which runs north and south in downtown Baltimore; several interesting neighborhoods are easily accessible from this corridor, including **Fells Point, Little Italy, Mt. Vernon, Canton, Charles Village,** and **Hampden.** Within a city this size, you could spend a lot of time visiting all kinds of sites. But this trip is intended to introduce you to the somewhat quirky, lesser-known, and exciting places beyond the typical harbor-area visits.

The history of the Baltimore & Ohio (B&O) Railroad is an integral part of how this country prospered, and it has at its heart Maryland locations from Baltimore to Ellicott City and beyond. While the railroad fostered commerce, literature fostered the imagination—and nothing in Baltimore's history is more mysterious and intriguing than the life and words of Edgar Allan Poe, who spent time on obscure streets in the city.

Mention the B&O Railroad and many immediately think of the space between Illinois and Atlantic Avenues on the Monopoly board. What they don't realize is that the B&O was a real railroad long before the game of Monopoly. The Baltimore and Ohio Railroad helped settle the eastern

United States by providing transportation for people and goods. In its wake, it left buildings, bridges, tracks, and railroad memorabilia and lore. It also brought extraordinary wealth to several of Baltimore's citizens, including philanthropists Johns Hopkins and George Peabody.

The B&O Railroad was founded in Baltimore, and had its headquarters here from 1827 to 1986. And the B&O was the largest employer in Baltimore for many decades. Sports lovers and railroad aficionados alike, as well as kids of any age, will get excited driving past **Oriole Park at Camden Yards,** which you do when you enter the city from the south. Like a giant amusement park, Oriole Park is a delightful re-creation of old-style baseball parks with modern touches and it is built on the site of an old B&O railroad yard. The long brick building on the first-base side used to be a railroad warehouse; it now houses offices, restaurants, and souvenir shops.

We're beginning our trip a few blocks farther up North Charles Street at the former **B&O Railroad Company Headquarters.** The building is on North Charles Street (where it intersects with Baltimore Street), only a 15-minute walk from the major league ballpark and even less from the Inner Harbor. Now an office building unconnected to the railroads, the B&O Building was once the distinguished headquarters of this important industrial giant.

Built in 1906 to replace the former headquarters, which was destroyed in the Great Baltimore Fire of 1904, the B&O building is a 13-story, steel-framed building. There are sculptures over the Charles Street entrance depicting the Roman mythological god Mercury (the symbol of commerce) and a figure representing the progress of industry (holding a torch and a locomotive).

In the summer of 2009, the top seven floors of the former B&O headquarters became a **Hotel Monaco,** with the boutique chain's usual perks of complimentary wine tastings and "companion goldfish" upon request. Hotel Monaco is a great place to stay if you plan to explore Baltimore for more than one day. The building also houses the **B&O American Brasserie** on its first floor (offices are located on floors two through six) with the hotel upstairs. If you don't want to stay at the hotel, you can have a drink at the bar or a meal in the restaurant in order to glimpse this spectacular building with its beaux-arts facade, Tiffany stained-glass windows, and marble interior.

After stopping at the B&O building to either check in at Hotel Monaco or to view the architecture, spend a few hours at the nearby **B&O Rail-**

Artistic details on the former B&O Railroad headquarters building

road Museum, site of the first railroad station in the country. One of the best railroad museums anywhere, the B&O museum has locomotives, exhibits, and a wonderful gift shop. To get there, drive north on Charles Street, make a left onto West Fayette Street before turning left onto Martin

Luther King Boulevard and making a quick right onto Lombard Street. Take another quick left onto Schroeder Street and a left onto West Pratt; the museum will be on your right.

After climbing aboard train cars and viewing exhibits at the museum, you'll no doubt be hungry. You're in luck, because Baltimore is a food lover's dream. Down-to-earth home-style cooking is everywhere, seafood is abundant, and Baltimore's style is generally casual and user friendly. There are, of course, many wonderful upscale restaurants. But for a sense of the real blue-collar workers' Baltimore, a walk around one of the markets, such as **Lexington Market,** is great fun. To get there from the railroad museum, go east on West Pratt for 0.5 mile and turn left onto South Paca Street for 0.4 mile. Turn left onto West Lexington Street and the market is on your right; there's an adjacent parking garage.

At Lexington Market, you can eat while walking around the food stalls, sampling the array of fantastic cooked foods that are interspersed with displays of raw ingredients. Crabcakes are de rigueur in Maryland—of course, along with rockfish, shrimp, and many other delicacies that come from the waterways in and around the state. But crabcakes are among the favorites. There'll be other places to try them later, but Lexington Market is one option that's been around since 1782. It's where locals have bought and eaten seafood literally for centuries. The neighborhood can be rough, so be careful. Another option for a Baltimore-style market is **Broadway Market** in Fells Point, where you'll go a little later and where it's generally safer to walk around.

Just as the former B&O Railroad warehouse is part of Oriole Park, Edgar Allan Poe is linked to sports in the city too. The king of the macabre and inventor of the modern mystery story, Edgar Allan Poe lived, worked, died, and was buried—not once, not twice, but three times—in Baltimore. The city's NFL football team, the Ravens, is named after Poe's poem "The Raven." The team even has three mascots, named—what else?—Edgar, Allan, and Poe.

Besides driving or walking past **M&T Stadium,** where the Ravens play when they're in town (and near Oriole Park), you can acknowledge Poe in other, more private ways. The best place to start paying homage to this literary genius is his grave site in the downtown Baltimore cemetery at **Westminster Church.** Though Poe was born in Boston and resided in Richmond and Philadelphia, it is his last resting place that most resonates with many of his fans. And many of Poe's fans come from around the world

BAWLMERESE

In Baltimore and the surrounding areas, residents have a unique way of pronouncing certain words. This "Baltimorese" is part dialect and part accent, and it is sometimes said to be a combination of Southern drawl and Northeastern style. Whatever its origin, it can be mildly confusing to visitors. Here's a mini-guide to help facilitate communication:

The city name: "Bawlmer" or "Bawldamore" (Baltimore)

What you drink out of a tap: "warder" or "worter" (water)

What you call a waitress or what she calls you: "Hon"

The kitchen floor is made up of this: "tahls" (tiles)

The state name: "Merlin" or "Murlin" (Maryland)

Kids play on it and adults walk on it: the "payment" (pavement)

The state capital: "nap-liss" (Annapolis)

Brides walk down an "ahl" (aisle)

"Down the oshin" (going to Ocean City for the weekend)

"Atzit" (that's it)

"Ca-meer" (come here)

"J'eet" (did you eat?)

---

to pay their respects at this cemetery, leaving flowers, pennies, Dove Bars, and even ballpoint pens (probably so he can continue writing). However, the cemetery is almost always quiet with few, if any, people around. Often you'll have the cemetery to yourself (except for those who are buried there), even on beautiful sunny days.

From Lexington Market, you can walk two blocks south on North Paca Street and turn right on West Fayette Street. The cemetery will be one block down on your left. If you want to drive, go west on Lexington and turn left on North Greene Street, left on West Baltimore Street, left on North Paca Street, and left on West Fayette Street. There's a parking garage on your left before you reach the cemetery, which is at the corner of Fayette and Greene Streets. Walking from Lexington Market is easier than driving because of the one-way streets, but either way you'll get there in 10 or 15 minutes.

You may wonder why Poe was buried three times. Like a character in one of his own tales, Poe, who died a pauper, was first buried in the wrong plot at the back of Westminster Church. This was corrected when he was moved to the right plot, also in back of the church. As his popularity grew after his death, Baltimore schoolchildren saved pennies until they had enough to move Poe to his third and more prominently situated grave. Now he is buried in the front of the cemetery, and his current monument resembles a penny with a bust of Poe's head. All three of the Poe grave sites are marked, so walk around the back of the church to see the first two.

Poor tormented Poe—buried three times. To make matters worse, his birth date is wrong on his current headstone. He was actually born on January 19, 1809, not January 20.

While you're wandering around the cemetery, you'll see that in addition to the remains of Poe, **Westminster Burying Ground** contains tombs of prominent political and military figures from Maryland's early history. Generals and patriots from the American Revolution and the War of 1812 are buried there. Adjacent to the **University of Maryland Law School** next door, the cemetery is actually a peaceful place to visit, even with catacombs beneath the church as well as supposed ghosts.

The Poe grave and surrounding cemetery and the church, with its early Gothic Revival architecture, are all intriguing. The cemetery grounds are gated and locked at night but remain open during daylight hours throughout the year. On a sunny day, a visit to the cemetery can be surprisingly relaxing—unless your imagination runs wild.

You can also visit the **Edgar Allan Poe House & Museum** only a few blocks away at 203 Amity Street, where the poet and master storyteller lived for three years. To get there, travel west and turn right onto Martin Luther King Jr. Boulevard. Then turn left on West Saratoga and left on North Amity. The narrow Amity Street residence, on the left, is typical of the downtown row houses for which Baltimore is well known—their marble steps recall a time that predated air-conditioning, when neighbors sat on their steps and talked to each other in the evenings. However, the Poe House is in a rough neighborhood and it's not always open, so you may want to just drive by and keep going.

On leaving the Poe House, you will drive south on North Amity and turn right onto West Lexington. Follow this with an immediate right onto North Schroeder Street and another right onto West Mulberry Street. Turn left onto Martin Luther King Jr. Blvd and right onto Druid Hill Avenue. Stay straight to go onto West Centre Street and then turn right onto Cathedral Street. The central branch of the **Enoch Pratt Free Library** will be on your right. The Enoch Pratt is a marvelous library worthy of its grand building. The library has a collection of odd Poe memorabilia, including a lock of his hair, a piece of his coffee pot, and several personal letters. There's an Edgar Allan Poe Room, but most of the Poe items are stored in the library's Special Collections section; if you're interested, the items can be brought out for viewing by special request.

At this point, you'll either be spooked by Poe and the characters he created or you'll be hungry and ready for dinner. Believe it or not, there's a restaurant named in honor of Poe's last poem, "Annabel Lee." First opened in 2007 in a building that dates back to 1907, the **Annabel Lee Tavern** is located in what is now the **Canton** neighborhood, but used to be considered a part of nearby Highlandtown. To get to the tavern from the Enoch Pratt Library, drive north on Cathedral and make a quick left onto West Mulberry for 1.4 miles and then a right onto North Wolfe Street for 0.7 mile. Turn left onto Eastern Avenue for 1.1 miles and then right onto South Clinton Street; the tavern will be on the left. Crabcakes are on the menu, along with sweet potato fries, burgers, crab and sweet corn quesadillas, and a chocolate Edgar Allan Pate for dessert. Specialty drinks are named appropriately, such as the Pit and the Pendulum. The Annabel Lee chicken salad is to die for (no Poe pun intended). Canton seems to have corner neighborhood watering holes at every street corner, but the Annabel Lee Tavern is worth traveling a distance to enjoy.

If you still want more Poe, you can visit **The Horse You Came In On Saloon** on Thames Street in the **Fells Point** neighborhood— the location of the docks where immigrants disembarked for centuries from points throughout Europe. The saloon is reputedly the tavern where Edgar Allen Poe imbibed his last bit of alcohol before his untimely demise. There's also a ghost walk you can take at dusk on Friday and Saturday evenings around the neighborhood. Or if you've had enough, you can check into the **Admiral Fell Inn** and take a ghost tour of the inn itself.

To get to Fells Point from the Annabel Lee Tavern, go north on South Clinton Street and turn left onto Eastern Avenue. Drive 1.3 miles past East Baltimore's Patterson Park and turn left onto South Broadway. The inn is 0.3 mile on the right. With so many bars lining the streets nearby, the rooms at Admiral Fell can be noisy, especially on weekends. If you want a view, ask for an outside room but if you want a quiet room, ask for an interior one.

To get to Fells Point directly from the Enoch Pratt library, go north on Cathedral and turn left on West Mulberry Street for 1.2 miles. Turn right on North Broadway and travel about a mile; the Admiral Fell Inn will be on the right. The inn is a quirky, boutique-style hotel located in a great neigh-

Pier in the Fells Point neighborhood

borhood for wandering and dining on local seafood. **John Steven Ltd.** is a favorite, especially for steamed spice shrimp. If you prefer mussels, go to **Bertha's.** Both are within a block or two of the inn. Wherever you dine, make sure to enjoy the delicious gelato at **Pitango Gelato** a few doors up the street from Admiral Fell Inn on South Broadway. You'll be interested to know that the streets are not composed of cobblestones, but rather of Belgian blocks, which were used as ballast in ships that landed here. The ballast was then used to pave the streets.

You can also take a short 1-mile walk to Baltimore's **Little Italy** neighborhood, where home-style Italian cooking has been thrilling locals and travelers for more than a century. To get from Fells Point to Little Italy, turn right on Thames Street from South Broadway, followed by another right on South Caroline and a left onto Lancaster Street. Turn right onto South Central Avenue and make a left onto Fawn Street. The restaurants in Little Italy are within a few blocks of each other.

Most, like **Sabatino's,** are longtime standards where your pasta is served *with* your entrée, not as a separate course. There's also the newer, more upscale **Aldo's,** which is also worth a visit. The owner of Aldo's is from Italy, of course, and he is a master craftsman who did all the hand-made woodwork in the restaurant himself, as well as doing the cooking while his wife and son run the front of the restaurant.

Another good choice is **Caesar's Den,** where the owner's wife makes excellent crabcakes. You can ask for a side of pasta with marinara sauce to accompany the crabcakes. Caesar's Den is at the corner of High and Styles Streets; across Styles Street from the restaurant is **Thomas D'Alesandro Jr. Park,** a tiny space with a full-sized bocce court (a sort of Italian bowling). It's a city court so you can play anytime; the locals recommend you bring your own bocce balls. D'Alesandro was mayor of Baltimore and his daughter, Nancy Pelosi, grew up in Baltimore and became Speaker of the U.S. House of Representatives.

After dinner, head back to Fells Point, which is one of Baltimore's most appealing nighttime neighborhoods. Its waterfront adds charm to the eclectic shops, taverns, and restaurants that line the streets. If you're staying at Admiral Fell Inn, you can spend the next morning having coffee by the docks and wandering around the stores. Be sure to look for the building facade used by director Barry Levinson to shoot the police station in the TV series *Homicide.*

After morning coffee, get back in your car and head uptown. From Admiral Fell Inn, turn left on Thames Street and make another left on South Broadway. Make a quick left onto Lancaster Street and an immediate right onto South Broadway. After a mile, turn left onto Orleans Street for 1.2 miles and right onto North Charles Street. You'll see Baltimore's **Washington Monument** at Mount Vernon Place, in the elegant **Mount Vernon** neighborhood. Although not as famous as the one in Washington, DC, Baltimore's monument was designed by the same architect, Robert Mills. Baltimore's 178-foot version predates the taller 555-foot one in Washington by 50 years. At one time, ships entering the Baltimore harbor could see the monument, as it's only 10 blocks from the Inner Harbor. But as the city skyline grew higher, the view of the monument from the waterfront became obscured.

Mount Vernon was home to many wealthy Baltimoreans in the late

Clock and sculpture over the entrance to Baltimore's Pennsylvania Train Station

1800s and early 1900s, and it is still a delightful neighborhood. The Garrett-Jacobs Mansion is there on Monument Street (which intersects Charles Street). The initial section of the mansion was built in 1872 by railroad tycoon John Garrett, who was the president of the B&O. The **Walters Art Gallery** and **Peabody Conservatory of Music** are also within a block of the Washington Monument.

Continuing up Charles Street, railroad aficionados will want to stop at **Baltimore's Pennsylvania Station** (nicknamed "Penn Station"). The old wooden benches in the old-fashioned waiting room are reminiscent of bygone eras when train travel was "the ultimate." Now serviced by Amtrak's East Coast routes, the station is an historic icon. There was talk of building a hotel on top of the station, but that project seems to have stalled. Plenty of temporary parking is located next door while you do a self-guided tour inside.

Two blocks away, at Charles and Chase Streets, is the former **Belvedere Hotel,** which has been converted into condominiums. However, it's worth a visit, for its infamous **Owl Bar** is still open on the first floor. Originally opened in 1903, the Owl Bar became a speakeasy during Prohibition. In-the-know travelers going from New York to Miami by train would stop in Baltimore, spend a few days at the Belvedere, and imbibe illegal booze at the bar. If the owl's eyes were blinking, it meant that the booze was there and the feds were not around. When his eyes were staring straight ahead and not blinking, patrons knew not to say anything. You can still drink or have a meal at the Owl Bar now—without fear of federal authorities trying to stop you.

Two miles farther north on Charles Street, you'll reach the **Homewood Campus** of **Johns Hopkins University (JHU)** in the **Charles Village** neighborhood. The university is named for a Quaker who made his fortune around the Civil War. Johns Hopkins fell in love with his cousin but his religion forbade him from marrying her. He never married anyone else and when he died with no wife or children, his immense wealth was given to Baltimore, first for his world-renowned namesake hospital and then the university.

OTHER BALTIMORE LOVE STORIES

Wallis Warfield Simpson, a Baltimore divorcée, captured the heart of King Edward VIII of England; unable to marry her while he was king (because she was a commoner and divorced), he abdicated his throne in 1936 so he could be with the woman he loved. The Baltimore house where Simpson once lived is still standing in the Bolton Hill neighborhood.

Love didn't conquer all with Jerome Bonaparte. Brother of Emperor Napoleon I of France, Jerome Bonaparte gave up Baltimore's Betsy Patterson so that he could become the King of Westphalia.

JHU is a really pretty 100-acre campus, with many traditional red brick and white marble Georgian-style buildings on meticulously landscaped grounds. Much of the wealth amassed by Johns Hopkins came from investing in the B&O Railroad, so there's a connection. Even without that connection, the campus of JHU is worth a visit.

After your visit to JHU, return to the waterfront at the Inner Harbor by taking St. Paul Street south (it runs parallel to Charles Street, which is one way north).

IN THE AREA

Accommodations

Admiral Fell Inn, 888 South Broadway, Baltimore, 21231. Call 410-539-2000 or 866-583-4162. A quirky, boutique-style hotel with 80 rooms, this inn is in the midst of the bars and restaurants of Fells Point. No two rooms at the inn are alike, which lends an air of anticipation to repeat stays. Web site: www.admiralfell.com.

Hotel Monaco Baltimore, 2 North Charles Street, Baltimore, 21201. Call 443-692-6170 or 800-546-7866. A luxury boutique hotel located on the top seven floors of Baltimore's historic B&O Railroad Headquarters Building. Beautiful architecture along with the Kimpton chain's branded amenities. There are 202 rooms and suites. Web site: www.monaco-baltimore.com.

Attractions and Recreation

Baltimore & Ohio (B&O) Railroad Museum, 901 West Pratt Street, Baltimore, 21223. Call 410-752-2490. This superb museum has a comprehensive collection of railroad cars and memorabilia. The gift shop has an enticing selection. Open daily. Web site: www.borail.org.

Basilica of the National Shrine of the Assumption of the Blessed Virgin Mary, 409 Cathedral Street, Baltimore, 21201. Call 410-727-3564. First Roman Catholic cathedral built in the United States (1806–1821). Web site: www.baltimorebasilica.org.

Bromo Seltzer Arts Tower, 21 South Eutaw Street, Baltimore, 21201. Call 443-874-3596. Modeled in 1911 after a famous landmark in Florence, Italy, the clock in the tower is still there with the words Bromo-Seltzer, but the giant bottle on top is long gone. The building is now used as space for artists. Web site: www.bromoseltzertower.com.

Edgar Allan Poe House & Museum, 203 North Amity Street, Baltimore, 21223. Call 410-396-7932. Edgar Allan Poe lived in this house for three years while he courted his cousin, whom he later married. Open Apr. through Nov., varying days and hours; it's best to call ahead. Web site: www.eapoe.org/balt/poehse.htm.

The Engineer's Club of Baltimore, at Garrett-Jacobs Mansion, 11 West Mount Vernon Place, Baltimore, 21201. Call 410-539-6914. Web site: www.engineersclubofbaltimore.org.

Enoch Pratt Free Library, Edgar Allan Poe Collection, Annex First Floor Corridor, 400 Cathedral Street, Baltimore, 21201. Call 410-396-5430. This majestic main branch of the library has a special collection of Edgar Allan Poe memorabilia. Web site: www.prattlibrary.org.

Homewood Campus of Johns Hopkins University, 3400 North Charles Street, Baltimore, 21218. Call 410-516-8000. One of Baltimore's most prestigious universities, the Homewood campus is the main location for this internationally renowned institution of higher learning. Web site: www.jhu.edu.

Jewish Museum of Maryland, 15 Lloyd Street, Baltimore, 21202. Call 410-732-6400. A large museum of Jewish culture and history with two adjacent historic synagogues, one of which is the third oldest in the nation. Web site: www.jewishmuseummd.org.

M&T Bank Stadium, 1101 Russell Street, Baltimore, 21230. Call 410-261-7283. This stadium is the home of the NFL's Baltimore Ravens football team. Web site: www.baltimoreravens.com.

Oriole Park at Camden Yards, 333 West Camden Street, Baltimore, 21201. Call 410-547-6234 or 888-848-2473. This old-style ballpark is home to the MLB Baltimore Orioles baseball team. Web site: www.orioles .com.

Pennsylvania Station, 1500 North Charles Street, Baltimore, 21201. Call 410-291-4165. This operating train station is the epitome of a long-gone era of travel. Amtrak operates the station as part of its East Coast Corridor system. Web site: www.amtrak.com.

Phoenix Shot Tower, 801 East Fayette Street, Baltimore, 21202. Call 410-605-2964. Molten lead was dropped 215 feet from this tower to make shot for guns from 1828 to 1892. Web site: www.carrollmuseums.org.

Walters Art Gallery, 600 North Charles Street, Baltimore, 21201. Call 410-547-9000. Free admission. Open Wed. through Sun. Web site: www .thewalters.org.

Washington Monument, 699 North Charles Street, Baltimore, 21201. Call 410-396-1049. This 178-foot monument was designed by architect Robert Mills, who also designed the Washington monument in Washington, DC. This obelisk was constructed between 1815 and 1829 and paid for by a lottery that raised $100,000. Open Wed. through Sun.

Westminster Hall and Burying Ground, 519 West Fayette Street, Baltimore, 21201. Call 410-706-2072. This is the cemetery where Edgar Allan Poe is buried. Administered by the University of Maryland Law School next door, the grounds are open during daylight hours. Occasional tours are held of the catacombs. Web site: www.westminsterhall.org.

Dining/Drinks

Aldo's Ristorante Italiano, 306 South High Street, Baltimore, 21202. Call 410-727-0700. Aldo's is newer and more expensive than many of the other restaurants in Little Italy, but it offers upscale Italian cuisine in a delightful setting worthy of special occasions. Open Mon. through Sat. for lunch and dinner. Web site: www.aldositaly.com.

Annabel Lee Tavern, 601 South Clinton Street, Baltimore, 21224. Call 410-522-2929. An eclectic and unpretentious tavern designed to celebrate Edgar Allan Poe. This is a visitor-friendly dining and drinking experience in the Canton neighborhood, which seems to have a bar on every corner. But this one is extra special. Excellent chicken salad and sweet potato fries. Open for dinner and drinks only, Mon. through Sat. Closed Sun. Web site: www.annabelleetavern.com.

Bertha's Restaurant & Bar, 734 South Broadway, Baltimore, 21231. Call 410-327-5795. The seafood at Bertha's, located in the heart of Fells Point, is legendary, especially the mussels. Open daily for lunch and dinner. Web site: www.berthas.com.

B&O American Brasserie, 2 North Charles Street, Baltimore, 21201. Call 443-692-6172. This restaurant is exciting because of its location in the landmark B&O Railroad Headquarters Building. A good place for drinks or dessert. Open for breakfast, lunch, and dinner daily. Web site: www.bandorestaurant.com.

Where Edgar Allan Poe may have had his last drink

Caesar's Den, 223 South High Street, Baltimore, 21202. Call 410-547-0820. One of Little Italy's long-standing restaurants, specializing in good seafood and pasta dishes. Open for lunch and dinner Mon. through Sat. Web site: www.caesarsden.com.

John Steven Ltd., 1800 Thames Street, Baltimore, 21231. Call 410-377-5561. Friendly tavern-like atmosphere and excellent steamed spice shrimp make this one of Fells Point's busiest restaurants. Open for lunch and dinner daily. Web site: www.johnstevenltd.com.

Lexington Market, 400 West Lexington Street, Baltimore, 21201. Call 410-685-6169. First opened in 1782, this is one of the markets in downtown Baltimore with fresh produce, seafood, and cooked foods for sale by many different vendors. Closed Sun. Web site: www.lexingtonmarket.com.

The Horse You Came In On Saloon, 1626 Thames Street, Baltimore, 21231. Call 410-327-8111. Rumored to be the bar where Edgar Allan Poe drank his last drink before he was found unconscious on the streets of Baltimore, leading to his untimely death at age 40. Web site: www.the horsebaltimore.com.

The Owl Bar, at the Belvedere, One East Chase Street (Charles & Chase Streets), Baltimore, 21202. Call 410-347-0888. Former Prohibition-era speakeasy that now serves legal alcohol and food for lunch and dinner. Upstairs on the 13th floor is another bar called the "The 13th Floor" that opens at 5 PM daily with great views of downtown Baltimore. Web site: www.theowlbar.com and www.MeetMeAtTheBelvedere.com.

Pitango Gelato, 802 South Broadway, Baltimore, 21231. Call 410-236-0741. This is a newer gelato company in the same location as one that was there previously, and it is just as good. Italian-style ice cream at its best. Web site: www.pitangogelato.com.

Sabatino's, 901 Fawn Street, Baltimore, 21202. Call 410-727-9414. One of Little Italy's longtime favorites. A tradition in Baltimore, with excellent pasta and seafood. Open daily for lunch and dinner. Web site: www .sabatinos.com.

Other Contacts

Baltimore Area Convention & Visitors Association, 100 Light Street, 12th Floor, Baltimore, 21202. Call 410-659-7300. Web site: www .baltimore.org.

Baltimore County Conference & Tourism, 44 West Chesapeake Avenue, Towson, 21204. Call 410-296-4886 or 800-570-2836. Web site: www.enjoybaltimorecounty.com.

Ed Kane's Water Taxis, 1735 Lancaster Street, Baltimore, 21231. Call 410-563-3901 or 800-658-8947. Web site: www.thewatertaxi.com.

Maryland Office of Tourism Development, 401 East Pratt Street, 14th Floor, Baltimore, 21202. Call 866-639-3526. Web site: www.visitmaryland .org.

Keeping score and watching the odds at Pimlico Race Track

CHAPTER

16

Hampden and the Horses

Thrilling in Hometown Baltimore

Estimated length: 60 miles
Estimated time: One day

Getting there: From downtown Baltimore, it is an easy drive to the Hampden neighborhood. Just travel north on Charles Street and turn left onto 33rd Street. Then turn right on Falls Road (MD 25) and right again onto West 36th Street.

Highlights: Preakness Stakes at Pimlico Race Track. State Fair at Timonium Fairgrounds. Lunch in Hampden at Café Hon or another local favorite. Ribs or crabs for dinner.

We'll start this day trip in **Hampden,** which is one of the great Baltimore neighborhoods that not everyone knows about. Hampden has a style and appeal all its own—part eccentric and part hometown Baltimore, with antiques and vintage "stuff," plus several good restaurants.

Originally a mill village in Baltimore County (which is a separate jurisdiction from Baltimore City), Hampden was annexed by the city in 1888. Yet its rural identity is still somewhat intact. To experience the character of Hampden, the four retail blocks on West 36th Street known as **the Avenue** should be your destination. The rest of the neighborhood consists of 19th-century-era residential row houses, many of which were originally built to house mill workers.

The Avenue looks like the center of a small town in the South, with its wide street and angled parking. Parking spaces are readily available, but be careful, for it is not clear that you actually have to pay for parking. Instead of obvious meters, there are inconspicuous machines that spit out paper receipts to be placed inside your windshield. Although the parking is a bargain, a traffic ticket is pricey if you get caught failing to pay. If you want to park a few blocks away from the retail section, on-street parking in the rest of the neighborhood is free.

Once you've parked your car, you'll be faced with a delightful choice of quirky shops and relatively inexpensive restaurants. The most famous place for food is **Café Hon,** named after the idiosyncratic expression "Hon," which is so typically Baltimore. The café even sponsors a **HonFest** for two days each June, during which attendees dress up in vintage '60s clothes and beehive hairdos. (The musical and film *Hairspray* are set in vintage Baltimore.)

Other than during the HonFest in June, and in December, when elaborate Christmas lights on 34th Street houses two blocks away draw huge crowds, Hampden is a sleepy, little-discovered, but charming place to shop for retro antiques, off-beat books, and artistic novelties.

Men and women can get a bargain haircut in an old-fashioned barber shop. Terrific works of art, including hand-painted furniture, can be purchased from the artist/owner of the shop **Hanging on a Whim.** And these are just a few of the many places where you'll find bargains and intriguing merchandise. Other places to eat, besides Hon Café, include **Holy Frijoles** and **Grill Art Café.**

After you've eaten and enjoyed some retail scavenger hunting in the stores, leave Hampden and drive north on Falls Road (MD 25) 1.5 miles, turning left onto Northern Parkway. If you follow this for 1.3 miles, you will reach **Pimlico Race Course,** site of the **Preakness**—the second leg of thoroughbred racing's Triple Crown in the **Park Heights** neighborhood. The Preakness is a once-a-year, not-to-be-missed event if you're in the area on the third Saturday of May. There is live racing at the track other days too, but it's the Preakness that is spectacular.

WIN, PLACE, OR SHOW

They're into the starting gate…those beautiful sleek thoroughbreds who are the athletes of the **Preakness Stakes.** The Preakness is one-third of the Triple Crown, and it occurs between the Kentucky Derby and the Belmont Stakes. The race takes place in Baltimore at Pimlico Race Track, the second-oldest track in the country (after Saratoga). No Triple Crown winner can claim his or her crown without first winning in Baltimore.

The winners are not limited to horses, jockeys, trainers, and owners. Everyone who attends the races on the third Saturday of May is a winner, whether you dress up and sit in the fancy sections or dress down and hang out on the grass in the infield.

The governor and other VIPs sit in boxes above the finish line, and when the weather cooperates, it's a great day in the sun. Black-Eyed Susans are the drink of choice—comparable to the Kentucky Derby's mint julep. Everyone has fun, especially those who properly handicap the horses and place bets on both favorites and long shots.

Between races at Pimlico

Exactas, trifectas, betting in "the box"… there's a whole vocabulary of betting that can bring good fortune for those astute enough, or lucky enough, to pick the winner or set of winners. Handicapping is the process by which people select horses for betting in individual races, with more care than merely selecting the horse's name. Such elements as genetics/pedigree, trainers, jockeys, and position at the gate are all figured into the equation.

Known as "Old Hilltop," Pimlico Race Track gets dressed up in all its finery for this annual event. The city starts the celebration a week in advance, with all kinds of entertainment and activities. On race day itself, men dress in slacks and sports jackets, women in sundresses and frilly hats. On the infield, it is perfectly acceptable to be in shorts or jeans, for at that part of the track, attendees are mostly milling around or sitting on the grass.

Food on Preakness Day includes the Black-Eyed Susans (official drink, the contents of which generally include vodka, rum, triple sec, orange juice, and pineapple juice, though the recipe changes) as well as crabcakes, Maryland fried chicken, coleslaw, and potato salad—traditional local dishes.

The race is alternately known as the Preakness Stakes and the "Run for the Black-Eyed Susans." The trophy is the silver Woodlawn Vase, a replica of which is presented to the winning owner. There's a Black-Eyed Susan blanket placed on the winning horse, and within minutes after the race is run, the race course weathervane is repainted with the number and colors of the winning jockey; the painting of the weathervane has been a Preakness tradition since 1909.

The first running of the Preakness Stakes for three-year-old horses was in 1873. Among the most famous three-year-olds are the Triple Crown winners— Affirmed, Seattle Slew, Secretariat, Citation, Assault, Count Fleet, Whirlaway, War Admiral, Omaha, Gallant Fox, and Sir Barton (first Triple Crown winner in 1919).

Reaching the finish line first is the important thing for this race that is $1\frac{3}{16}$ miles long. Other races are held during Preakness day, but there is only one Preakness Stakes.

While this sporting event is only one day each year, horse racing and breeding are a part of the state culture and economy. While it's not Kentucky, Maryland is considered a thoroughbred state and there are many beautiful horse farms.

If you are doing this trip another day, or you choose not to watch this ultimate day of horse racing at Pimlico, after leaving Hampden, drive north on Falls Road (MD 25) 1.5 miles, turn left onto Northern Parkway, and immediately exit onto the Jones Falls Expressway (I-83) heading north. You'll soon reach the Baltimore Beltway (I-695), an interstate that encircles Baltimore City and part of the "other Baltimore"—Baltimore County.

Take I-695 west toward Towson and exit back onto I-83 north. Your destination is **Timonium,** where the Maryland State Fairgrounds are located. On most weekends, there are craft shows and other activities held at the fairgrounds. From I-83 take exit 17, Padonia Road, east to York Road. Go south on York Road to the fairgrounds entrance on the right.

For 11 days at the end of summer (late August to early September), you can go to the Maryland State Fair, which is held here.

The State Fair has rides, concerts, livestock and horse shows, demonstrations, exhibit halls, an amusement park, rides, and a racetrack with horse racing on weekends.

Aside from the big events, Maryland is a state of hometowns, with residents who trace their local roots back many generations. No traditions are more sacred than culinary ones, especially in Maryland. Food provides a way of bonding, of social interaction, of friendship extended and enjoyed.

No aspect of life in Maryland is more reflective of this than the love of fresh seafood emanating from the Chesapeake Bay and the plentiful rivers throughout the state. No specific food source is more beloved than the ubiquitous Maryland blue crab. Even in years when the crab is endangered, imports allow for locals to enjoy hardshell steamed crabs with Old Bay Seasoning, crabcakes, crab soup, and other variations.

There are numerous crab houses in the Baltimore area where you can enjoy the Old Bay–covered crustaceans, steamed and messy, but absolutely delicious. If you ask locals, they will be opinionated about the best places. A few that are frequently mentioned in the Baltimore area include **Reter's Crab House** in Reisterstown and **Gunning's** in Hanover, but there are many other great crab houses in the Baltimore area and throughout the state. As long as you have good company, fresh crabs, and plenty of time, you're bound to have a good time.

EATING STEAMED HARDSHELL CRABS

A true Maryland phenomenon, eating hardshell crabs around a table covered with brown wrapping paper or newspaper, pitchers of beer at hand, is one of the best summertime ways of entertaining and being entertained. To outsiders, those hardshell crabs may look hard to eat. But it's really easy once you get the hang of it. There are minor variations, with proponents of one technique or

Wonderful, delectable steamed hardshell crabs

another usually insistent that their way is the best. However, the basics are the same. Here is a short primer:

Wear casual and comfortable clothes. Cover any cuts on your fingers with Band-Aids so the spices don't burn your skin. Have a sip of beer or iced tea, and pick up your first crab.

Break off the claws, sucking on the crabmeat at the ends. Put the claws aside for later.

Remove the crab's apron with your fingers. Break it off and discard. Pull the hard outer shell off with one hand while holding the base of the crab in your other hand. Remove the "devil" meat from the top of the crab's body. Discard the devil.

Break the body of the crab in half and gently pick at the crabmeat from inside the body, avoiding the softer shells surrounding the delicious and succulent crabmeat. You can eat the "mustard" with the crab if you like; some people like it, some don't. The same with the squiggly white "stuff." But the bulk of the goodies are inside the two halves with the hard-core crabmeat. Enjoy.

Pick up a second crab and start the process all over again. You'll eat the claws last (after you've had your fill of crabs) with the aid of a wooden mallet, possibly a metal nutcracker, and a small, sharp knife.

NOTE: Marylanders can generally eat a dozen or so crabs at one sitting, and are often satisfied eating just the crabs themselves. A crab feast (whether with two or two hundred participants) is a long, involved social event with great conversation and storytelling. Visitors should be satisfied eating a few crabs on their first outing; there's usually sweet white Maryland corn on the cob, crab soup, and crabcakes around too.

Crab Smarts: Eating crabs is a leisurely activity. Those prone to enjoying long conversations with a cold beer or soda and eating with their fingers are best suited for these marathon sessions. If a particular crab doesn't taste right, discard it immediately even if it's expensive; a bad crab can make you sick and it's not worth taking a chance. Tell stories. Have fun.

IN THE AREA

Attractions and Recreation

Hanging on a Whim, 828 West 36th Street, Baltimore, 21211. Call 410-467-3233. Web site: www.hangingonawhim.com.

Maryland State Fairgrounds, 2200 York Road, Timonium, 21093. Call 410-252-0200. Located at the intersection of York and Timonium Roads. Largest event is the State Fair held for 11 days in late August and early September. Web site: www.marylandstatefair.com.

Pimlico Race Course, Northern Parkway and Pimlico Road (or Hayward and Winner Avenues), Baltimore, 21215. Call 410-542-9400. This thoroughbred racetrack is old and well loved. Its claim to fame is the annual running of the Preakness Stakes, the second leg of horseracing's Triple Crown. Web site: www.pimlico.com and www.preakness.com.

Towson Town Center, 825 Dulaney Valley Road, Baltimore 21204. Call 410-494-8800. Huge destination shopping mall. Web site: www.towsontowncenter.com.

Dining/Drinks

Café Hon, 1002 West 36th Street, Baltimore, 21211. Call 410-243-1230. Located on "the Avenue" in Hampden, this down-home-style café gets lots of attention because of its crabcakes, cinnamon buns, and pork chops. The best dessert by far is the homemade bread pudding, well worth a try even if it's not your favorite dessert, for Café Hon's version is terrific. Service is sometimes surly, which is unlike Baltimore generally. Open daily. Web site: www.cafehon.com.

Corner Stable, 9942 York Road, Cockeysville, 21030. Call 410-666-8722. If you can't make it to Pimlico for the Preakness, you can watch the race here while eating delicious barbecued ribs. Skip the crabcakes, though, for they are not good. Web site: www.cornerstable.com.

Grill Art Café, 1011 West 36th Street, Baltimore, 21211. Call 410-366-2005. Open daily.

Gunning's Seafood Restaurant, 7304 Parkway Drive, Hanover, 21076. Call 410-712-9404. Open daily. Web site: www.gunningsonline.com.

Holy Frijoles, 908 West 36th Street, Baltimore, 21211. Call 410-235-2326. Open daily for lunch and dinner.

Jilly's Sports Bar & Grill, 1012 Reisterstown Road (MD 140), Pikesville, 21208. Call 410-653-0610. Excellent crabcakes. Open daily for lunch, dinner, and drinks. Web site: www.jillyssportsbar.com.

Jumbo Seafood, 48 East Sudbrook Lane, Pikesville, 21208. Call 410-602-1441. Delicious Chinese food in a small strip center. Lamb with cilantro is excellent. Closed Mon.

Reter's Crab House, 509 Main Street, Reisterstown, 21136. Call 410-526-3300. Web site: www.reterscrabhouse.com.

Tark's Grill, 2360 Joppa Road, Suite 116, Lutherville, 21093. Call 410-583-8205. Excellent crabcakes. Web site: www.tarksgrill.com.

Towson Diner, 718 York Road, Towson, 21204. Call 410-321-0407. Excellent crabcakes and rockfish stuffed with crabmeat. Open daily 24 hours.

Other Contacts

Baltimore Area Convention & Visitors Association, 100 Light Street, 12th Floor, Baltimore, 21202. Call 410-659-7300. Web site: www .baltimore.org.

Baltimore County Conference & Tourism, 44 West Chesapeake Avenue, Towson, 21204. Call 410-296-4886 or 800-570-2836. Web site: www.enjoybaltimorecounty.com.

Concord Point Lighthouse overlooking the waters around Havre de Grace

CHAPTER

17

Havre de Grace Waterfront

Touring the Lower Susquehanna River

Estimated length: 120 miles
Estimated time: Day trip or overnight

Getting there: From Baltimore, take I-95 north to exit 89. Follow MD 155 east and stay in the left lane. Where the street comes to a dead end, turn right onto Juniata Street. At the first traffic light, turn left onto Ostego. At the next dead end, turn right onto Union Avenue. Go to the end and turn left onto Commerce Street (road name will change to Market Street). Follow around the bend in the road; there is parking on Giles Street.

Highlights: Charming waterfront town. Historic lighthouse. Decoy museum. Excellent seafood restaurants. Homemade ice cream. A ride on a working skipjack. Minor league baseball game at Ripken Stadium. Topiary gardens.

Named for the Susquehannock Indians—the most powerful tribe of the upper Chesapeake Bay when the European settlers first arrived—the **Susquehanna River** was the location of a prosperous fur trade with the Indians. The Susquehanna is the largest non-navigable river in the United States; every minute it pours 19 million gallons of fresh water into the Chesapeake Bay.

Situated right where the Susquehanna meets the bay, **Havre de Grace** is a beautiful, somewhat underappreciated town and river port; its French name means "harbor of grace." There's a grace to the town that fits the name, and the people are friendly and open to visitors.

Havre de Grace is a classy waterfront town with excellent restaurants, luxurious bed & breakfasts, a working skipjack upon which you can take rides, and a lighthouse next to a short pedestrian boardwalk. The town has lots of charm and even a touch of sophistication.

For this trip, start out by visiting the **Havre de Grace Decoy Museum,** with its expansive exhibits of wildfowl decoys used for hunting in this Upper Chesapeake Bay area.

Next go to see the **Concord Point Lighthouse.** Decommissioned by the Coast Guard in 1975, it is kept up by citizens who formed a group to ensure the upkeep of the historic lighthouse, which was constructed in 1827. The lighthouse is situated in a beautiful spot.

After you've seen the lighthouse, have lunch at one of the restaurants in town. They are all good, but **Laurrapin Grille** is superb with absolutely delicious crab gazpacho and crab/lobster cakes. The food at this gastropub is not to be missed.

In season, you should also arrange to take a ride on the skipjack *Martha Lewis.* The *Martha Lewis* is a real working skipjack, one of the few left in the state, and it is still used for oystering in the bay when it is not ferrying visitors around the Havre de Grace waterfront. Watching the team hoist the sails and maneuver the ship is fascinating. Your job is to stay out of the way while enjoying the experience.

Afterward, you may want homemade ice cream from **Bomboy's Homemade Ice Cream** in town or you can drive out into the country to **Broom's Bloom Dairy.** To reach Broom's, take I-95 to exit 80. Head toward Bel Air on MD 543. This is a winding road with the dairy on the left about 4 miles from the exit.

While you're out in the country exploring, you might head north to see **Ladew Topiary Gardens,** which is always a treat with its many comical and entertaining figures designed out of plants. The artistic ingenuity of the landscapers is amazing. While there, consider taking a tour of the manor house; if you do, keep your eye out for a secret entranceway hidden behind a panel of books in the library.

To reach Ladew, take I-95 to exit 74; follow MD 152 north to a left on MD 146. After seeing Ladew, you can head back to I-695 (Baltimore Belt-

Characters carved from trees and bushes at the Ladew Topiary Gardens

way) on MD 146 south. Or you can head back into Havre de Grace for dinner. If you decide to stay over, the **Vandiver Inn** bed & breakfast is particularly pleasant.

Or you can head to nearby Ripken Stadium and catch a baseball game if the Aberdeen IronBirds are playing on their home field. To reach the stadium from I-95, take exit 85 to MD 22 west. Turn right at the light onto Long Drive.

Sports lovers are fond of baseball great Cal Ripken Jr., who grew up in the Havre de Grace-Aberdeen area before joining the Baltimore Orioles. He's returned to start this minor league team and built a wonderful stadium in the process; the stadium is only about 15 minutes from Havre de Grace.

Marylanders love their baseball. And while the Baltimore Orioles and Washington Nationals are around for many to cheer on, the minor league teams around the state are worth attention too. If you're going to be in and around Havre de Grace during baseball season, you would enjoy catching one of the IronBirds games. The stadium is smaller, of course, than the ones used in the majors, but it's pretty upscale; the level of play is good, and the food options—always important at a baseball game—are pretty good themselves.

Between the restaurants, the baseball, the topiary plants, and the maritime history, you may just decide to make visiting Havre de Grace a regular habit. It's one of those intriguing little places hiding in plain sight.

IN THE AREA

Accommodations

Vandiver Inn, 301 South Union Avenue, Havre de Grace, 21078. Call 410-939-5200 or 800-245-1665. There are 8 rooms with fireplaces and antique beds, plus 10 more rooms in the two houses next door. All rooms have private baths. Old-fashioned decor, but it's fresh and clean, plus there's wireless Internet. A scrumptious and generous breakfast is included in the price. Web site: www.vandiverinn.com.

Spencer Silver Mansion, 200 South Union Avenue, Havre de Grace, 21078. Call 410-939-1097 or 800-780-1485. The Victorian house was built in 1896 and has four guest rooms, two of which share a bath and two of which have private baths, plus a stone carriage house out back with room enough for two. Web site: www.spencersilvermansion.com.

Attractions and Recreation

Concord Point Lighthouse, Lafayette & Concord Streets, Havre de Grace, 21078. Call 410-939-3213. The oldest continuously operating lighthouse in the state, authorized to help with hazardous navigation where the Susquehanna River meets the Chesapeake Bay. Illumination has gone from whale oil lamps with tin reflectors to more modern lenses. Lighthouse was automated in 1920 and decommissioned in 1975 by the Coast Guard. Open weekends Apr. through Oct. Free admission. Web site: www.concordpointlighthouse.com.

Havre de Grace Decoy Museum, 215 Giles Street, Havre de Grace, 21078. Call 410-939-3739. Open daily. Admission fee. Web site: www .decoymuseum.com.

Havre de Grace Maritime Museum, 100 Lafayette Street, Havre de Grace, 21078. Call 410-939-4800. Web site: www.hdgmaritimemuseum .org.

Ladew Topiary Gardens, 3535 Jarretsville Pike, Monkton, 21111. Call 410-557-9466. Sculpted topiary trees and bushes, a beautiful estate, and a formal rose garden. Mansion is also pretty fabulous. Open daily Apr. through Oct. Web site: www.ladewgardens.com.

Ripken Stadium, 873 Long Drive, Aberdeen, 21001. Call 410-297-9292. Home stadium for the Aberdeen IronBirds, a farm team for the Baltimore Orioles. Named for former Oriole superstar Cal Ripken Jr., who is the principal owner. Web site: www.ironbirdsbaseball.com.

Skipjack Martha Lewis, 121 North Union Avenue, Havre de Grace, 21078. Call 410-939-4078. In season, this sailing vessel and working oyster dredge provides a good way to get out on the water and take a look at the town from that vantage point. The skipjack is the official state-designated boat. Sails Apr. through Oct. Web site: www.chesapeakeheritage.org.

Broom's Bloom Dairy, where you can get homemade ice cream

Dining/Drinks

Bomboy's Homemade Candy & Ice Cream, 322 and 329 Market Street, Havre de Grace, 21078. Call 410-939-2924 or 877-266-2697. Candy shop is on one side of the street; ice cream shop on the other. Web site: www.bomboyscandy.com.

Broom's Bloom Dairy, 1700 South Fountain Green Road (MD 543), Bel Air, 21015. Call 410-399-2697. Beautiful farm with homemade ice cream. Web site: www.bbdairy.com.

Laurrapin Grille, 209 North Washington Street, Havre de Grace, 21078. Call 410-939-4956. Local artists have painted a wall mural, the top of the bar, and the tables. Bar atmosphere with sophisticated food. Open for lunch and dinner; closed Mon. Web site: www.laurrapin.com.

MacGregor's Restaurant & Tavern, 331 St. John Street, Havre de Grace, 21078. Call 410-939-3003 or 800-300-6319. Delicious food in a

casual setting. Open 365 days a year. Web site: www.macgregorsrestaurant
.com.

Tidewater Grille, 300 Franklin Street, Havre de Grace, 21078. Call 410-939-3313. On the waterfront along the Susquehanna River. Cajun Jamba-laya is especially good; they warn you that it's spicy, but it's not so much. Great views of the railroad bridge over the water. Web site: www.the tidewatergrille.com.

Other Contacts

Harford County Office of Tourism, 220 South Main Street, Bel Air, 21014. Call 410-638-3327. Web site: www.harfordmd.com.

Havre De Grace Office of Tourism, 450 Pennington Avenue, Havre de Grace, 21078. Call 410-939-2100 or 800-851-7756. Web site: www.hdg tourism.com.

Edge of the Northeast River

CHAPTER

18

Chesapeake City, North East, and Elkton

Wandering around the Upper Chesapeake Bay

Estimated length: 140 miles
Estimated time: Overnight

Getting there: From Baltimore, take I-95 north to exit 100. Take MD 272 south approximately 13.5 miles to Elk Neck State Park.

Highlights: Historic village of Chesapeake City along the Chesapeake & Delaware (C&D) Canal. Town of North East with antiques and gift shops. Turkey Point Lighthouse in Elk Neck State Park. Two covered bridges. Homemade ice cream. Beautiful scenery and horse farms. The Historic Little Wedding Chapel—an East Coast version of quickie Las Vegas chapels.

Driving on I-95 between Baltimore and New York City, most drivers stay on the interstate, merely stopping at rest stops for gasoline, restrooms, and snacks. At the Chesapeake House, one of the rest stops near the Maryland–Delaware border, many travelers may not know about the intriguing small towns that lie not far off the main road.

Three of these towns are **North East** (yes, that's the name of the town near the **Northeast River,** with the river name one word and the town name two), **Chesapeake City,** and **Elkton.** On this trip, you'll explore these towns and the surrounding countryside. You won't be disappointed by these hidden treasures, unknown to even many Marylanders. Although this part of the state on the upper Chesapeake Bay is heavily trafficked, most travelers on I-95 are intent upon their destination rather than the journey. When you slow down and get off the interstate, there are delightful sights and tastes to enjoy.

This trip is full of many activities and if you plan to fit everything in, you'll want to stay overnight. There are several good bed & breakfasts in Chesapeake City and there's also an excellent B&B in North East—**Woodland Gardens,** which overlooks the Northeast River.

If you're a lighthouse enthusiast, you won't want to miss **Turkey Point Lighthouse,** which is at the head of the Chesapeake Bay in **Elk Neck State Park.** Drive to the end of the road you're on (MD 272) and park your car at the lighthouse parking lot. You'll need to walk about 1 mile (actually 0.8 mile if you take the shortest route) to reach the lighthouse.

The park itself is 2,200 acres, with miles of trails and a beach. The lighthouse is situated on a bluff 100 feet above the bay, so on clear days you can enjoy the view knowing that five rivers flow together to form the top, or the upper portion, of the Chesapeake Bay: the Susquehanna, Northeast, Elk, Bohemia, and Sassafras Rivers. The Elk River is right below the bluff and it flows into the C&D Canal a little bit north of your location.

Turkey Point is a short tower as lighthouses go, but it stands on the 100-foot bluff, which makes it one of the highest lights off the water in the bay. Four of the 10 known keepers were women—this was unusual in the heyday of lighthouse keepers, though 3 of the women were wives who succeeded their husbands, which was not uncommon.

The beacon from Turkey Point Lighthouse was visible for 13 miles, which helped guide boaters into the mouth of the then newly completed C&D Canal. Decommissioned by the Coast Guard in April 2000, the light was reactivated in 2002 by a nonprofit group. Even if you're not a fan of lighthouses, it's an excuse to hike in a pretty state park along a trail through the woods with cliffs nearby overlooking the Northeast River and the Chesapeake Bay. On weekends from late April to mid-November, the lighthouse is open so you can climb the spiral staircase to the top to take in the view from that vantage point. There are no admission or parking fees.

From the state park, you're only about 10 miles from North East, a wonderful small town. Go north on MD 272 until you get to the light at Cecil Avenue. Turn left and then make another left onto Main Street.

In North East, there are antiques and gift shops, plus a few great restaurants, especially **Woody's Crab House.** Woody's has superb crabcakes and you can also get steamed hardshell crabs served on the standard brown-paper-covered tables with wooden knockers for cracking claws. This is a cheerful place where you'll enjoy sitting for hours eating the steamed crabs. There are also all-you-can-eat roasted peanuts in the shells, friendly service, and instruction for anyone who is new to eating steamed crabs.

The town is small but interesting, with shops like **Where Butterflies Bloom** and the **5 & 10 Antique Market,** which was previously a five-and-ten and, for the first half of the 20th century, the Hotel Cecil.

After lunch, you have several options. You can get on MD 272 north and go visit **Kilby Cream** for some homemade ice cream and a chance to see a few animals on a farm. If you choose this option, cross over I-95 and turn left onto Joseph Biggs Highway (MD 274). Turn left at Barnes Corner Road, right onto Hopewell Road, and right onto Strohmaier Lane. Kilby Cream will be on the right.

Another option for the afternoon is to head north to **Fair Hill Natural Resource Management Area (NRMA).** This area is similar to but not exactly like a state park. It is managed by state park rangers, but its purpose is the protection of natural resources, whereas state parks are generally geared to recreation. However, there is much to enjoy at Fair Hill. There are trails for hiking or horseback riding. If you don't have your own horse, you can rent one at Fair Hill Stables. Every year for one day—the Saturday of Memorial Day weekend—there are also steeplechase races, with betting, at Fair Hill.

To reach Fair Hill from North East, turn left on any downtown street and left again onto Mauldin Avenue, which merges with MD 272 north. On the way, you'll pass **Gilpin Falls Covered Bridge.** It is no longer used for car travel (there's a newer bridge right next to it where you will drive), but you can park your car by the side of the road to take a photo or to walk through the 119-foot-long bridge built in 1860 to span Northeast Creek. Its purpose was to allow horses, wagons, and riders to reach a mill that was located on the other side.

From North East, drive about 8 miles to MD 273. Turn right (east) onto Telegraph Road (MD 273) and drive about 5 miles. Turn right into Entrance 3, turn left onto Ranger Skinner Road, and left onto Training Center Road, crossing over MD 273. Turn right onto Tawes Drive, and on your right is the NRMA office, where you can purchase a trail map and ask the rangers for information and directions.

Within Fair Hill is one of only six wooden covered bridges still standing in Maryland. It is the **Fox Catcher Covered Bridge** and it spans a stream. The bridge was originally built in 1860 but suffered damage over the years and was reconstructed in 1992. Fair Hill also has four original **Mason-Dixon Line** 1-mile markers. These are the markers installed by Jeremiah Dixon and Charles Mason in the 1760s to settle a land dispute between William Penn and Lord Baltimore; the line became the border between Maryland and Pennsylvania.

Fair Hill borders Pennsylvania and few of the original markers are left intact and in place, so it's significant that four of these markers are on Fair Hill property. You can talk to the rangers to get directions for hiking on the property to see at least one of the markers, as they are not easy to find. At least one is exposed in a field, another in a briar patch.

After you've spent time at Fair Hill, if you're staying at Woodland Gardens Bed & Breakfast go back to the town of North East. Continue through town on MD 272 south and after 1 mile, bear right onto Hances Point Road. The B&B is in a beautiful contemporary house with 10 acres of grounds on the Northeast River, and the hosts are delightful.

You can plan to have a casual dinner at **Rivershack** across the river in Charlestown, and your hosts can help you arrange transportation by boat. By land from North East, turn left onto Cecil Avenue (MD 7) and bear left at MD 267. At the stop sign turn left and you'll come to the Rivershack. Alternately, you can have dinner at one of the restaurants in downtown North East.

From Woodland Gardens, you can plan to visit Chesapeake City in the morning, after you are served a wonderful breakfast outside on the patio.

If you're staying at a B&B in Chesapeake City instead, head west from Entrance #3 at Fair Hill on MD 273 and make a left onto Singerly Road (MD 213) going south toward Chesapeake City. Cross US 40 and then in 6 miles you'll cross the **Chesapeake City Bridge**. At the bottom of the bridge, turn right and continue going right after the yield sign. This will curve you around underneath the bridge. At the first stop sign, turn left onto George Street and then turn right at Fourth Street. Make a left onto Bohemia Avenue and look for an available on-street visitor parking space.

Chesapeake City is in the shadow of the Chesapeake City Bridge silhouetted high above it and spanning the **C&D Canal**. It is a teeny town, with more of a "village" feel, and was once actually called the Village of Bohemia. It evolved and grew when the canal was built.

Much activity in Chesapeake City revolves around the C&D Canal.

The C&D (short for the Chesapeake and Delaware) Canal is an important, vibrant canal. It was dug by hand with intensive manual labor and first opened for business in 1829. In 1927, it was dredged out further and turned into a sea-level waterway, eliminating the need for all the locks.

In 1942, a ship accident destroyed the bridge that connected the two sides of Chesapeake City. For the next seven years, the only way to cross the canal was by ferry. Then the current high-level bridge was opened in 1949, but it sort of bypassed the tiny town. With this newer bridge in place, if you want to go to Chesapeake City, you actually have to drive in a circle underneath the bridge. However, the canal has prospered. And the town is still there, tucked under the bridge, or at least in its shadow, with great scenic views of the canal and the oceangoing cargo ships that transport their goods through the canal.

The C&D Canal is the busiest canal in the United States and the third-busiest canal in the world, just after the Panama and Suez canals. It provides a shortcut from the Delaware River to the Chesapeake Bay and the Port of Baltimore. For boats on the eastern seaboard, this is a big deal, though on some days you can be there without seeing any shipping vessels, just recreational ones. On other days, the canal will be busy with seagoing traffic. The canal is operated by the U.S. Army Corps of Engineers.

As in other parts of Maryland, growth was dependent upon the economical transportation of goods. With the Delaware River and the Chesapeake Bay separated by only a narrow strip of land, a waterway connecting them made sense in reducing by almost 300 miles the water route between Philadelphia and Baltimore. Still, the construction of such a project in the early 1800s was a huge and expensive undertaking. Once it was built, teams of mules and horses towed freight barges through the canal's system of locks. Steam pumps were implemented as the years progressed. Eventually it became clear that a wider and deeper canal was needed.

In 1919, the federal government bought the canal and designated it as part of the Intracoastal Waterway. The number of ships and the amount of cargo kept increasing, so further expansion was needed. All locks except one were removed as the canal was converted to a sea-level operation. Now approximately 40 percent of the ship traffic to and from the Port of Baltimore travels through the canal.

Horse country surrounds Chesapeake City. To see some of the farms, take MD 286 west out of town, turn right onto Old Telegraph Road and right again onto MD 310 (Cayotes Corner Road). If you time it right, you'll see lovely thoroughbred horses for there are several breeders in this area. Turn right onto MD 213 and you will come back into Chesapeake City just before the bridge.

Next you'll head toward the town of Elkton by getting back onto MD 213 north. Cross over the intersection of MD 40 and keep going straight on MD 213. At Main Street, turn right. Once on Main Street, go straight through the light at North Street and look for the **Historic Little Wedding Chapel** on the right. There are some free on-street parking spaces that are good for two hours.

Elkton was once a popular marriage destination, for there was no waiting period to obtain a marriage license. It is the county seat, so the courthouse is here, along with lawyers' and bail bond offices. After seeing the wedding chapel, stop in at **Main Street Café** for bagels and coffee if it's early, or stop in for a drink and bar food if it's later in the day. You can also get an old-fashioned milk shake or other food at the nostalgic **Lyons Pharmacy Luncheonette.** All of this is right on Main Street.

The town of Elkton was named for its location on the Elk River, at the "Head of Elk." The name was shortened to Elktown and then in 1787 to Elkton. In the 1700s, Elkton was an important trading location due to the gristmills and ports located nearby.

From the 1920s to 1940s, many couples took the train from New York, Philadelphia, and other areas to get married at the wedding chapels that flourished in Elkton because it was in a jurisdiction with no waiting period. Famous people who did this included President Eisenhower's son John, basketball star Charles Barkley Jr., baseball star Willie Mays, and singer Billie Holiday.

Elkton was the first county seat on the way into the state from the north, so it became known as the "Marriage Capital." The Historic Little Wedding Chapel is the last of the wedding chapels from that era that still exists. Couples can still get married in the Chapel (after a waiting period), as well as at the courthouse across the street.

To return to Baltimore, drive past the Historic Little Wedding Chapel on Main Street and take the next right. Then turn right again onto Howard Street. At MD 213 (when you can't go any further), turn right and then turn right again onto MD 279. Watch for the signs to get onto I-95 south.

IN THE AREA

Accommodations

Blue Max Inn Bed & Breakfast, 300 Bohemia Avenue, Chesapeake City, 21915. Call 410-885-2781 or 877-725-8362. Within walking distance from the center of town and the waterfront. Web site: www.bluemaxinn.com.

Inn at the Canal, 104 Bohemia Avenue, Chesapeake City, 21915. Call 410-885-5995. Lovely Victorian B&B with views of the C&D Canal. Web site: www.innatthecanal.com.

Woodland Gardens Bed & Breakfast, 555 Hances Point Road, North East, 21901. Call 410-287-2479. Gorgeous B&B on the Northeast River with a pier from which you can arrange a boat ride with advance notice. Super-relaxed setting. Web site: www.woodlandgardensbnb.com.

Calf enjoying the sunlight at Kilby Cream farm

Attractions and Recreation

Chesapeake & Delaware (C&D) Canal Museum, 815 Bethel Road, Chesapeake City, 21915. Call 410-885-5621. The canal itself is more impressive than the museum, but if you want to learn about the canal, you can stop in to see the exhibits. The views from the museum site are interesting, with the Chesapeake City Bridge overshadowing the horizon and signs showing that the spot is equidistant from the Chesapeake Bay and Delaware River (15 miles in each direction). Museum is open Mon. through Fri. Free admission.

Elk Neck State Park, 4395 Turkey Point Road (MD 272), North East, 21901. Call 410-287-5333. Turkey Point Lighthouse, trails, camping, fishing, swimming, and picnicking. Web site: www.dnr.state.md.us.

Fair Hill Natural Resources Management Area, 300 Tawes Drive (MD 273, Entrance 3), Elkton, 21921. Call 410-398-1246. Operated by the Maryland Department of National Resources, State Forest & Park Service. Horseback riding, hiking trails, four Mason-Dixon Line mile markers, a covered bridge, fishing, and a training center for racehorses. Open year-round. Web site: www.dnr.maryland.gov.

Gilpin's Falls Covered Bridge, alongside the road on MD 272, north of I-95, exit 100, North East, 21901.

Historic Little Wedding Chapel, 142 East Main Street, Elkton, 21921. Call 410-398-3640. Web site: www.historiclittleweddingchapel.com.

Turkey Point Lighthouse, in Elk Neck State Park. Call 410-287-5333. Open Sat. and Sun., late Apr. through mid-Nov. Web site: www.tpls.org.

Vulcan's Rest Fibers, 106 George Street, Chesapeake City, 21915. Call 410-885-2890. Great shop for knitters and other do-it-yourselfers, plus beautiful handmade items for sale. Open daily. Web site: www.vulcanrest.com.

Where Butterflies Bloom, 102 South Main Street, North East, 21901. Call 410-287-2975. A charming gift shop that sells a variety of treasures, especially unique handmade soaps.

5 & 10 Antique Market, 115 South Main Street, North East, 21901. Call 410-287-8318. Open daily.

Dining/Drinks

The Bayard House Restaurant, 11 Bohemia Avenue, Chesapeake City, 21915. Call 410-885-5040 or 877-582-4049. Good food, good service, and waterfront views of the C&D Canal. Mostly seafood and steaks. Lunch and dinner daily. Web site: www.bayardhouse.com.

Chesapeake Inn Deck, 605 Second Street, Chesapeake City, 21915. Call 410-885-2040. Great place to sit outside and enjoy good food and a cheerful atmosphere. More casual than the restaurant inside. Live music. Web site: www.chesapeakeinn.com.

Kilby Cream, 129 Strohmaier Lane, Rising Sun, 21911. Call 410-658-8874. Milk from the cows on this farm is turned into the homemade ice cream sold on-site. You can also visit (at no charge) some animals (calves, ponies, and pygmy goats) in the backyard, a playground for young children, and a corn maze in the fall. Open varied days throughout the year depending on the season. Closed Jan. Web site: www.kilbycream.net.

Lyons Pharmacy Luncheonette, 107 East Main Street, Elkton, 21921. Call 410-398-2820. A perfect place to get an old-fashioned milk shake or other snack food. Affords a glimpse into the past, when luncheonettes were plentiful.

Main Street Café, 126 East Main Street, Elkton, 21921. Call 410-620-5100. Closed Sun. Web site: www.bagelscafeandmore.com.

Rivershack at the Wellwood, 523 Water Street, Charlestown, 21914. Call 410-287-6666. Steamed hardshell crabs, fried chicken, and pizza are the staples here, along with draft and bottled beer, plus other casual fun food. Web site: www.wellwoodclub.com.

Steak & Main, 107 South Main Street, North East, 21901. Call 410-287-3512. Upscale steakhouse; one section is less formal than the other. Open daily for lunch and dinner. Web site: www.mysteakandmain.com.

Woody's Crab House, 29 South Main Street, North East, 21901. Call 410-287-3541. A deservedly popular place to enjoy excellent crabcakes and steamed hardshell crabs. Located in the heart of downtown North East, with shops all around. Lunch and dinner daily. Web site: www .woodyscrabhouse.com.

Other Contacts

Cecil County Tourism, 68 Heather Lane, Suite 43, Perryville, 21903. Call 410-996-6299 or 800-232-4595. Web site: www.seececil.org.

Elkton Chamber of Commerce and Alliance, Web site: www.elkton alliance.org.

Historic Chesapeake City. Call 410-885-2415. Web site: www .chesapeakecity.com.

Town of North East, 106 South Main Street, P.O. Box 528, North East, 21901. Call 410-287-5801. Web site: www.northeastmd.org.